LAWS OF THE GAME

Translated by Robert and Rita Kimber

LAWS of THE GAME

How the Principles of Nature Govern Chance

Manfred Eigen
and Ruthild Winkler

Princeton University Press

Princeton, New Jersey

Published by Princeton University Press, 41 William Street,
Princeton, New Jersey 08540

Library of Congress Cataloging-in-Publication Data

Eigen, M. Manfred, 1927–
[Spiel. English]
Laws of the game: how the principles of nature govern chance /
Manfred Eigen and Ruthild Winkler.
p.   cm.—(Princeton science library)
Translation of: Das Spiel.
Originally published: 1st American ed. New York: Knopf:
Distributed by Random House, 1981.
Includes bibliographical references (p.   ) and index.
ISBN 0–691–02566–5 (paperback)
1. Science—Philosophy. 2. Chance. 3. Scientific recreations.
4. Games. I. Winkler, Ruthild, 1934–   . II. Title. III. Series.
[Q175.E3713   1993]
501—dc20   92–16084

First Princeton Paperback printing, 1993

Princeton University Press books are printed on acid-free paper
and meet the guidelines for permanence and durability of the
Committee on Production Guidelines for Book Longevity of the
Council on Library Resources

10 9 8 7 6 5 4 3 2 1

Printed in the United States of America

Contents

Translators' Note   *vii*

Acknowledgments   *ix*

Foreword   *xi*

Foreword to the English Edition   *xv*

ONE The Taming of Chance   *1*

1 The Origin of Play   *3*

2 Games People Play   *6*

3 Microcosm—Macrocosm   *19*

4 Statistical Bead Games   *30*

5 Darwin and Molecules   *49*

TWO Games in Time and Space   *67*

6 Structure, Pattern, Shape   *69*

7 Symmetry   *103*

8 Metamorphoses of Order   *131*

Contents   vi

THREE The Limits of the Game—The Limits of Humanity *173*

   9 The Parable of the Physicists *175*

  10 Of Self-Reproducing Automata and Thinking
      Machines *178*

  11 "From One Make Ten . . ." *199*

  12 Limited Space and Resources *216*

  13 From Ecosystem to Industrial Society *236*

FOUR In the Realm of Ideas *249*

  14 Popper's Three Worlds *251*

  15 From Symbol to Language *259*

  16 Memory and Complex Reality *283*

  17 The Art of Asking the Right Question *298*

  18 Playing with Beauty *306*

    List of References *331*

    Index *339*

TRANSLATORS' NOTE

In *Laws of the Game,* the authors often illustrate a point by referring to games like chess or checkers or backgammon, which are familiar to English-speaking readers. Occasionally, however, they discuss games that are not well known in this country. In these few cases, we have substituted games that illustrate the same points the authors want to make but that will also be instantly recognizable to an American audience. In Chapter 2.1, for example, we have used tic-tac-toe in place of the original's "nine men's morris."

Similarly, where the original illustrates a principle with an example drawn from the German language, as in the discussion of entropy in Chapter 8.3, we have adapted the example to English.

The authors' system of annotation, which we have retained in the translation, differs from that generally used in this country. Whereas superscript note indicators in the text normally refer to consecutively numbered bibliographical notes that give page numbers as well as bibliographical data for the work in question, here each superscript represents instead the number of a given title in the "List of References" at the end of the book. The reference is only to the work as a whole, not to any specific passage within it; and the same note number is used each time that particular work is cited. Quotations from Hesse's *Glass Bead Game,* for instance, appear in Chapters 1 and 18.3; and both quotations are followed by the number 3, which refers the reader to that title in the List of References.

Robert Kimber
Rita Kimber

Acknowledgments

Many friends and colleagues have helped us in writing this book. We would like to take this opportunity to thank them all.

Material for illustrations was kindly provided by Max Perutz, Konrad Lorenz (who drew some sketches for us in the course of a discussion), and Benno Hess (who conducted some experiments expressly for us). The sources from which other illustrations were taken are identified in the captions. Peter Richter, together with Nancy Williams and Bernd Morgeneyer, simulated a number of bead games on the computer. From the large number of results, we were able to abstract typical patterns.

We received expert advice in areas with which we are less familiar, from Mogens Schou (lithium therapy), Theodor Wolpers (Joyce translation), Walther Zimmerli (Old Testament), Adel Sidarus (Arabic studies), and Karl Vötterle (musicology). We also received valuable aid from colleagues whose fields of physics, chemistry, and biology are closely related to ours, particularly from Otto Creutzfeldt, Hans Frauenfelder, Ernst Ruch, Reinhard Schlögl, Peter Schuster, and Klaus Weber. We remember with pleasure many of the apposite anecdotes told us by our friends Shneior Lifson, David Nachmannsohn, and Charles Weissmann, and we have incorporated a number of these into the text. Francis Otto Schmitt's workshops in the Neurosciences Research Program at M.I.T., held over several years, have provided us with many insights into the subject of "information processing in the central nervous system."

Last but not least, we would like to thank our friends who read through the manuscript before it went to press and called our attention to unclear passages. These readers were Hans Frauenfelder, Hans Herloff Inhoffen, Klaus Oswatitsch, Peter Schuster, and Renate Böhme, and Frieder Eggers, Wilhelm Foerst, Peter Markl, Peter Richter, and Hans Rössner.

M.E.   R.W.-O.

Foreword

The lay reader interested in the natural sciences is exposed to a constantly rising flood of information. He may often feel himself cast in the role of a judge before whom different authors appear like competing litigants, each of whom hopes to be found in the right. If these authors are sufficiently clever in presenting their cases, the reader may be swayed by each in turn. But then a reviewer appears on the scene and objects that each cannot be as right as the other, particularly when one author's claims are the exact opposite of another's. The reader has no choice but to find the reviewer right, too.

This familiar anecdote, which we have slightly adapted to suit our purposes, contains a moral: Everyone could in fact be right if each one did not insist on being the only one to be right.

Everything that happens in our world resembles a vast game in which nothing is determined in advance but the rules, and only the rules are open to objective understanding. The game itself is not identical with either its rules or with the sequence of chance happenings that determine the course of play. It is neither the one nor the other because it is both at once. It has as many aspects as we project onto it in the form of questions.

We see this game as a natural phenomenon that, in its dichotomy of chance and necessity, underlies all events. Our interpretation of games and play thus goes far beyond Huizinga's, which considers play only in its relation to human behavior. In applying the concept of play to art, we come much closer to Adorno's views, which clearly disagree with Huizinga's identification of play and art.

The considerations that gave rise to this book go back to the molecular theory of evolution worked out a few years ago. It also draws on the models developed in connection with that theory, models for simulating phenomena like equilibrium, selection, and growth, all of which are based on natural laws. Although this book makes frequent use of examples drawn from the field of biology, its scope is much broader and includes general scientific, philosophical, sociological, and aesthetic perspectives. Our main purpose is to show play in its symbolic nature and in all its metamorphoses. We also make use of its different aspects to illuminate both our scientific understanding of the world and various philosophical positions. We are aware that this brings us into conflict with those who attribute validity to one and only one aspect.

The actual writing of the book was preceded by innumerable talks and discussions with friends and colleagues during our Engadin winter seminars, on hikes and ski tours in the mountains, and, occasionally, over a bottle of wine. We first thought a dialogue would be the appropriate form for our exchange of ideas, but we soon abandoned this concept. The classical dialogues are without exception artificially constructed. Simplicio and Salviati, for example, are both mouthpieces for Galileo. We finally decided that we would both take turns going over all the material point by point. We have, of course, ventured into areas in which we are only dilettantes. We hope the reader will forgive us this in the light of our efforts to uncover parallels and so to emphasize the unity of nature and mind.

We do not want to go any further into the substance of the book here than is necessary to make clear why we chose the title we did. Similarly, the introductions to the individual chapters are meant to lead the reader into the material treated, not to summarize it.

Chance and law (or principles) are the basic elements of games. The subtitle of the book suggests the interaction between them. And here we should add that it is the *consequences* of chance that are subject to regulation. Only if there is a large number of individual events does the element of chance come under the control of statistical law. This law is at work, for instance, when chance fluctuations undergo self-regulation in equilibrium or when they are amplified in the evolutionary process. The origin of the genetic code, the development of languages by which we transmit our thoughts, the intellectual play of artistic imagination are all based on the same fundamental principles of evolution, even though the results of individual games are determined by the caprice and variability of chance.

The manifestations of the game that matter plays in time and space and the effects this game has on human beings are the themes of the central chapters in the book. Much as we may agree with Jacques Monod in his view of molecular biology, we differ from him greatly in the conclusions he draws from it for mankind and society. We see Monod's demand for an "existential attitude toward life and society" as an animistic inflation of the role of "chance." Such a demand neglects the complementary role played by natural law. Criticism of the dialectical overemphasis on "necessity"—a criticism we find fully justified—should not move us to deny completely its obvious influence.

We are in complete agreement with Monod's statement that ethics and knowledge cannot be divorced from each other. For us, however, this view does not condemn the great religions but rather sets them a new task. As we have occasion to point out in the present work, the natural sciences yield no proof of God's existence, but neither do they claim that men "can do without faith in God."

The unity of nature is expressed more in the laws governing the relationships between structures than in the structures themselves. Taking this "motto" as a point of departure for Part IV, we examine a number of problems that have recently come more and more under the purview of the natural sciences and that are still far from being solved. In these considerations, we are mainly concerned with the mechanisms of cognizance. In terms of the elementary processes involved, these mechanisms always amount to a falsification in Karl

Popper's sense. Still, we cannot fully accept the views of this great epistemological logician, particularly when he claims: "There is no induction." Good enough! But there are different mechanisms of falsification that vary in their degree of inductive adaptation.

One of our main concerns in this book lies at the heart of Part III. There we raise the still unsolved question of the hubris of human knowledge. Friedrich Dürrenmatt's categorical imperative "What concerns all can only be worked out by all" is as relevant for the physicist whose knowledge can unleash the forces of nature as it is for the biologist who is now in a position to manipulate genetic material and influence psychic behavior with drugs. And our appeal here is also directed at the economist and the politician, who have the task of creating, maintaining, and protecting the conditions necessary for a worthwhile human existence.

Pessimistic predictions based on simple extrapolation are widespread today. We will not be able to solve the problem of limited resources by frugality alone. A raw material that will be used up in about fifty years at present rates of consumption might last a hundred years if we cut down on its use. (Such a reduction might, of course, threaten the stability of our economy.) In other words, the catastrophe could be delayed for another fifty years. Would it not be more sensible to concentrate on ensuring unlimited resources, perhaps by reusing our raw materials in a recycling economy and by developing sufficiently productive sources of energy? However we solve this problem, the practicability of our solution will depend on stopping population growth, perhaps even on reducing the world's population.

We must be clear about one thing: The human being is not a fluke of nature, nor does nature automatically guarantee his survival. The human being is one player in a huge game, the outcome of which is, for him, uncertain. He has to make full use of his capabilities to hold his own as a player and not become a plaything of chance.

Manfred Eigen / Ruthild Winkler-Oswatitsch
Göttingen, September 1975

FOREWORD TO THE ENGLISH EDITION

Five years have passed since the German publication of *Laws of the Game*. In that time, human understanding of nature has increased enormously. This is especially true for this book's central area of concern: the field of molecular biology. Some of the techniques described here as "brand new" have now become standard procedures; some of the discoveries called "very recent" have been acknowledged with Nobel Prizes, others have been superseded by more important and deeper insights. Our aim in *Laws of the Game* was to present abstract formulations of laws based on observed regularities in nature. These formulations, we believe, are timeless even though some of their possible applications in reality remain controversial.

Readers will decide for themselves whether or not to accept the lesson of our experience, and of this book, that questions need to be looked at from many different angles. In reality, there are few problems that can be solved by a simple yes or no. The challenge of true problems is unique. Their solutions demand a thoughtful and balanced consideration of all the "pros" and "cons" involved.

Manfred Eigen / Ruthild Winkler-Oswatitsch
Göttingen, August 1980

ONE

The Taming
of Chance

Research in physics has shown beyond the
shadow of a doubt that in the overwhelming
majority of phenomena whose regularity
and invariability have led to the
formulation of the postulate of causality,
the common element underlying the
consistency observed is chance.

Erwin Schrödinger, "What Is a Natural Law?"
(Inaugural Address delivered at the University of
Zurich, December 1922)

1

The Origin
of Play

Play is a natural phenomenon that has guided the course of the world from its beginnings. It is evident in the shaping of matter, in the organization of matter into living structures, and in the social behavior of human beings.

The history of play goes back to the beginnings of time. The energy released in the "big bang" set everything in motion, set matter whirling in a maelstrom of activity that would never cease. The forces of order sought to bring this process under control, to tame chance. The result was not the rigid order of a crystal but the order of life. From the outset, chance has been the essential counterpart of the ordering forces.

Chance and rules are the elements that underlie games and play. Play began among the elementary particles, atoms, and molecules, and now our brain cells carry it on. Human beings did not invent play, but it is "play and only play that makes man complete."[1]

All our capabilities arise from play. First, there is the play of limbs and muscles. The aimless grasping and kicking of an infant develop into carefully coordinated movements. Then there is the play of our senses. Playful curiosity sends us in search of profound knowledge. From play with colors, shapes, and sounds emerge immortal works of art. The first expressions of love take the form of play: the secret exchange of glances, dancing, the interplay of thoughts and emotions, the yielding of partners to each other. In Sanskrit, the union of lovers is called *kridaratnam,* "the jewel of games."[2]

Every game has its rules that set it apart from the surrounding world of reality and establish its own standards of value. Anyone who wants to "play" has to follow those rules. In parlor games, rules established before the game begins determine the course the game will take and define the scale of values by which it is played. But the effects of a chance occurrence can change the constellation of the game and set it running in a totally new direction. This is how life initiated its first games, and this is how our thoughts and ideas continue to play those games.

> From the terrifying realm of elemental forces and the holy realm of laws arises the aesthetic impulse, which creates a third, carefree realm. This is the realm of play and appearance. In it, man is freed from the bonds of all circumstance and liberated from compulsion, moral as well as physical.[1]

It is our intention in this book to trace the interplay of chance and law that has come down to us from the farthest reaches of time and found its most perfect expression in the inexhaustible creativity of our ideas. We hope to translate Hermann Hesse's symbol of the glass bead game back into reality. Hesse described this game as one that incorporated all the values of our culture.

> All the insights, noble thoughts, and works of art that the human race has produced in its creative eras, all that subsequent periods of scholarly study have reduced to concepts and converted into intellectual property—on all this immense body of intellectual values the Glass Bead Game player plays like the organist on an organ. And this organ has attained an almost unimaginable perfection; its manuals and pedals range over the entire

intellectual cosmos; its stops are almost beyond number. Theoretically this instrument is capable of reproducing in the Game the entire intellectual content of the universe.[3]

There is something mysterious about glass beads. The reflection and refraction of light make them glow, and as we explore the idea of play in this book, Hesse's glass beads will take on a new life. Part of their symbolic significance derives from the fact that they can exemplify the constant transformations and metamorphoses we observe all around us: atoms evolve into crystals, molecules into genes, living cells into cognitive beings, letters into words, notes into harmonies. The course the bead game takes is determined by the roll of the dice, yet at the same time it is also influenced by the rules of the game, just as chance in nature is subject to the laws of physics. The dice and the rules of the game—these are our symbols for chance and natural law.

2

Games People Play

In games of chance, in games of strategy, and in games involving both chance and strategy, the course that play takes on any given occasion will be "historically" unique because of the large number of possible choices involved. The sequence of moves constantly opens new directions the game can follow along the branches of a decision tree. The arbitrary course that play takes at each fork of the decision tree depends on the chance roll of the dice as well as on each player's ignorance of his opponent's strategy. Game theory determines criteria for an optimal strategy in any given situation, and these criteria also hold for dealing with economic and political problems.

2.1 GOOD LUCK AND BAD

In games, we speak of good luck when chance is kind to us, of bad luck when it is not. With these terms, we impose values on a basically neutral concept.

Lotteries are games of chance in their purest form. Their appeal lies in the prospect of winning a prize, whether a large sum of money or a tube of toothpaste. Even the most modest prize is often enough to fire enthusiasm and raise a gambler's passion to fever pitch. But a lottery without a prize would be no fun at all. Chance would be nothing but chance, neither good luck nor bad.

It is worth noting about lotteries that we perceive them only as bringers of *good* luck. We know, of course, that our chances of winning are extremely slim, and because we are always prepared to lose, a win is all the more pleasant a surprise—all the more a stroke of good luck—when it comes.

But there are other games that revolve around bad luck. Parcheesi is an example. The point is to put your opponent in a disadvantageous situation, and the setbacks the players suffer determine the outcome of the game. What could be more annoying than being sent back to the start when you are just a few moves short of your goal?

Why do children like Parcheesi so much? Because their parents have convinced them that overcoming adversity in a game is fun and is good preparation for dealing with difficult situations later in life? Or is the pleasure of foiling someone else's plans so great that we gladly accept the risk of sometimes being foiled ourselves?

Games that are determined purely by chance are just as boring as those whose limited number of variations make the outcome predictable. In tic-tac-toe, for example, an experienced player can always at least draw provided he does not make any mistakes.

But what does "not making any mistakes" mean? To avoid mistakes, a player has to anticipate all the possible moves his opponent can make—and this often involves thinking many moves ahead—before he can decide what the "correct" move is. The mental process a player has to go through in choosing his next move can be represented by a decision tree with so many branches that it is impossible to survey them all at once (see Figure 1). Even if there were only two alternatives to choose from at each fork (see Figure 2), a player would still be faced with a dense "thicket" of possible moves, and the prospect of finding one's way through such a tangle would hardly strike most players as much fun.

Figure 1. The delta of the Colorado River on the Gulf of California resembles a huge "decision tree." The innumerable branches in the coastal area are caused by the movement of tidal ebb and flow. (Scale 1:20,000. This illustration is used with the permission of the Aero Service Corporation, Philadelphia, Pa.)

Most of us are more likely to act on impulse; but, as we all know, our impulses are far from reliable.

If you had the choice of taking a million dollars in a lump sum or of receiving one penny on the first day of a month, two on the second, four on the third, and so on throughout the month, which would you

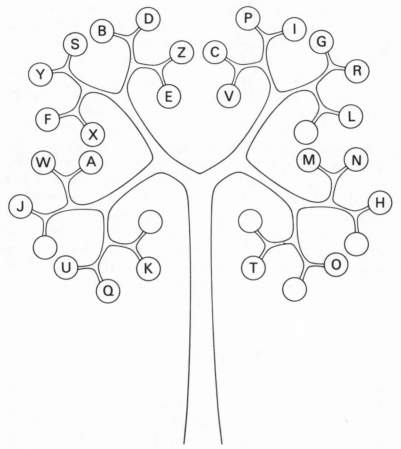

Figure 2. "Decision Tree." The principle of the decision tree is used here to represent the teletypewriter code. Any letter is reached by way of five binary decisions. At each fork, either the right (0) or left (1) branch is chosen. The letter *R,* for instance, is symbolized by the code word 01010.

choose? Forced to make a quick decision, most of us would instinctively opt for the single payment of a million dollars. After all, that is quite a handy sum. Why take chances?

But the second offer is in fact a far better one, particularly if the month in question has thirty-one days. In that case, the sum would amount to $10,737,418.24.

Businessmen often have to make similar choices between immediate cash returns and long-term income. If they rely on their impulses, the consequences can be disastrous. For them, it is essential to consider, either in their heads or on paper, the effects of every possible choice the decision tree presents.

Let us examine more closely a game involving a finite decision tree. The best way to do this is to play the game, and because the game is competitive, we will need an opponent to play it. This game, which probably originated in China, was rediscovered and analyzed by the American mathematician Charles L. Bouton early in this century. Bouton called the game "Nim," deriving this name from an Old English word that means "take." Bouton worked out rules for ensuring a win every time, but we would spoil the game for you if we revealed them here. (Anyone who wants to play "Nim" to win should consult Walter R. Fuchs's book *Moderne Denkmaschinen* [*Modern Thinking Machines*].[4] In it, Fuchs describes the "nimitron," a kind of computer that some bright English students developed and that assures victory to the player who uses it.)

2.2 GAME THEORY...

We are not so much interested in *how* the game "Nim" can be won as we are in the fact that if the player who makes the first move knows what he is doing, he will inevitably win. But if he does not know the trick on which victory depends, he will have to run through all the possible choices the decision tree presents before he can make a move. The knowledgeable player, however, will be spared this trouble because mathematicians have long since worked out a precise analysis of this game. In the terminology of game theorists, "Nim" is a finite,

Table 1. B E A D   G A M E   ''N I M''

A random but not too small number of beads is arbitrarily divided into several groups and placed on a board in separate rows (see Figure 3). The players take turns at removing the beads from the board. At each turn, they may take one or more beads from any one row; or, indeed, they may take a whole row. Whoever removes the last bead is the winner. Each player therefore tries to create a situation that forces his opponent to remove the next-to-last bead or row of beads from the board.

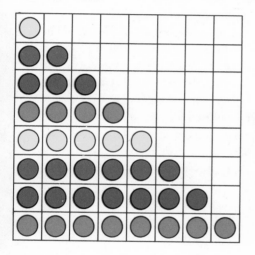

Figure 3. "Nim." One possible layout for a game of "Nim" is illustrated here. Eight groups of colored beads are divided into eight horizontal rows. At each turn, a player may take as many beads as he likes from the board, provided that he takes them from one row only.

two-person, zero-sum game with perfect information and optimal strategy. Translated into everyday language, this forbidding description means simply:

1. that the game is played by two players,
2. that it ends after a finite number of moves have been made,
3. that there are always a winner and a loser, i.e., that the sum of what is won and what is lost always amounts to zero, and
4. that there is one clear strategy that insures victory to the player who makes the first move, regardless of what his opponent may do.

These conditions provide an ideal basis for a theoretical discussion because they allow us to formulate a foolproof strategy that will make the course of play for a game of this kind completely predictable every time this strategy is applied.

But in the area of economic problems, the conditions are not so clear-cut as a rule, and it is usually impossible to formulate an optimal strategy in advance. Such a strategy will depend—and in some cases it will depend decisively—on a number of external factors that are more or less unpredictable. In situations of this kind, all we can do is act on certain assumptions about probability that promise the best possible results under the given conditions.

Game theory was developed to deal with problems of this kind. The founder of game theory was the Hungarian mathematician John von Neumann, who worked and taught at the Institute for Advanced Study in Princeton from 1933 until his death in 1957. Game theory was originally designed for solving economic problems, and the classic work on the subject by von Neumann and Oskar Morgenstern was entitled *Theory of Games and Economic Behavior.*[5] This theory has proved fruitful not only in economics but also in sociology, politics, and military strategy; and it has also given rise to the field of futurology.

We will attempt to outline, in general terms at least, the essence of this theory, the ramifications of which mathematicians have as yet by no means fully explored. (For a detailed introductory study, see Morton Davis, *Game Theory: A Nontechnical Introduction* [Basic Books, 1970].)[5a] Some readers will know the children's game of stone-paper-scissors. In this game, the two players simultaneously put out their hands, making a sign that represents either stone, paper, or scissors. Scissors cut paper; paper wraps stone; stone smashes scissors. This is a two-person, zero-sum game, but it lacks both complete information and optimal strategy. And, unless the players put a time limit on it, it is not "finite" either. In several respects, then, it differs from the game of "Nim" we have just described. Since neither of the players can predict his opponent's decisions, there can be no optimal strategy. Every round is a complete game with a winner and a loser.

If we want to represent a game and its possible courses in rational terms, we have to list the conceivable alternatives or strategies open to both players and evaluate the outcomes that result from all possible combinations. The formal device used to do this is called a "pay-off matrix." This matrix is nothing more than a table that helps each player decide on his optimal strategy.

For stone-paper-scissors, the pay-off matrix looks like this:

PLAYER 1

		stone	scissors	paper
	stone	0	−	+
PLAYER 2	scissors	+	0	−
	paper	−	+	0

A plus sign means a win for the first player and, because this is a zero-sum game, a loss for the second player. A minus indicates a loss for the first player and, of course, a win for the second. Zero represents a draw, which occurs if both players happen to make the same choice.

There would be little point in setting up a matrix like this for zero-sum games with perfect information because the course that games of this kind will follow is predetermined. It is always possible for one player—in principle, at any rate—to choose a strategy that will guarantee him a win no matter what his opponent does. There is, of course, a world of difference between theoretically perfect information and information that is actually available. Games like chess, checkers, and go are, in principle, determined, but in reality their outcome is always unpredictable. Perfect information for chess is so complex that not even our largest computer could be sure of a win. For these games, too, it would be possible to set up a pay-off matrix for comparing and evaluating different strategies, provided we dealt with only a limited number of moves. But the huge number of variations possible within even a relatively few moves would soon make such a matrix too complicated to be of any practical use.

Most card games are by nature games with imperfect information. For them, a pay-off matrix shows the best course of action or how a player should respond to the opposition's moves. Should he pass? Should he take a given trick or let his opponent have it? In cards, we usually have some clues to our opponents' intentions because we know that they are trying to win, too, and that they will consequently try to foil all our actions. But in games like stone-paper-scissors, we have no idea what our opponent will do, and a pay-off matrix here cannot help us choose a strategy. All it does is provide us with a concise presentation of the possible outcomes of the game.

The aim of game theory is to achieve the optimal gain in any given situation. By means of a pay-off matrix, it determines how specific strategies can best be applied to realize the desired gain. The minimax theorem, which John von Neumann proved mathematically in 1928, is crucial here. This theorem states that in a finite, two-person, zero-sum game, an average return, designated by the letter V, is always assured for one of the players, assuming that both players play sensibly. More precisely:

1. There is a strategy for player I that will protect this return; against this strategy, nothing that player II can do will prevent player I from getting an average win of V. Therefore, *player I will not settle for anything less than V.*
2. There is a strategy for player II that will guarantee that he will lose no more than an average value of V; that is, *player I can be prevented from getting any more than V.*
3. By assumption, the game is zero-sum. What player I gains, player II must lose. Since player II wishes to minimize his losses, *player II is motivated to limit player I's average return to V.*[5a]

The third statement is listed separately from the second because it applies only to zero-sum games. Player I's gains affect player II only if they come out of player II's pocket. But in non-zero-sum games one player's gains are not necessarily the other player's losses, and neither player will, as a rule, try to thwart his opponent out of sheer spite. In a game of this kind, what both players desire could be described as a prearranged distribution of the gain, and "informed" players will be prepared to accept this distribution from the outset.

In mathematical terms, a zero-sum game represents an optimization problem that can easily be solved if the side conditions are clearly defined. But in most optimization problems the side conditions are variable and can be bounded only in terms of minima and maxima. The business world provides ready examples: Competitors cannot set their prices above established price ceilings, but they may sell their products at lower prices. They may have limited warehouse space or production capacity. Conditions affecting production and sales are variable, and if an optimal gain is to be achieved, these variables must be balanced out against each other. In calculating an optimum, we therefore cannot hope for the same precision we can get in calculating maxima and minima, on a clear curve for which the side conditions are fixed and constant.

Almost all economic thinking involves optimization problems. Indeed, an economy cannot function at all unless optimal solutions are sought for its problems. It is obvious that competition for markets bears some resemblance to a zero-sum game. But it is equally obvious that the ideal conditions for a zero-sum game are rarely met in the field of economics, and that the competition is rarely limited to two opponents.

The recommendations made by the Club of Rome were based on considerations of this kind. In an overpopulated world, the gains of one group often represent the losses of another. What raises the standard of living may well diminish the quality of life.

2.3   ...AND HUMAN BEHAVIOR

But we should not deceive ourselves into thinking that we will fully grasp reality with the aid of a theory that almost always assumes ideal conditions. In his foreword to Morton Davis's book on game theory, Oskar Morgenstern writes:

The reader of this book will be impressed by the immense complexity of the social world and see for himself how complicated ultimately a theory will be that explains it—a theory compared to which even the present-day difficult theories in the physical sciences will pale.[5a]

That does not mean, of course, that a theory capable of elucidating the social behavior of human beings will have a special place outside the framework of the natural sciences. Man is the product of an evolution reflecting universal natural laws. The explanation for human behavior, complex as that behavior may be, will ultimately be found in the basic principles of science. We cannot predict, of course, what levels of complexity our understanding will reach. Theories are based on abstraction. They abstract what is regular and readily reproducible from reality and present it in an idealized form, valid only under certain assumptions and boundary conditions. Game theory is no different.

The most unpredictable factor in the practical application of game theory is the actual behavior of the players. Can they be depended on to follow the dictates of reason as defined by game theory, or isn't it more likely that they will try to upset their opponents' calculations by whatever possible means? Any truly comprehensive game theory would therefore have to include a thorough understanding of the human psyche.

Why are card games so popular? Unlike chess, checkers, or go, they are rarely "open." Card players cannot know the exact distribution of the cards and have no sure way of predicting their opponents' actions. By bluffing, a player can often make the best of a poor hand. The rules of most card games are quite simple and require no great intellectual effort to master. All that is usually involved is a simple scale of value: the high card or high hand wins. Psychological acumen is more decisive than abstract thought. Some players are so adept at reading minds that they can anticipate their opponents' moves even in games like stone-paper-scissors.

Imagine how boring skat would be if it were played with open cards. The game would consist of two equally uninteresting phases: (a) the hands themselves, which are products of pure chance, and

(b) the playing of the cards, which, in this case, follows a predetermined course dictated by the rules of the game. The whole appeal of the game would be lost, because it is the combination of chance and rules that makes it interesting. When the players keep their cards to themselves, the element of chance contained in the hands continues to exert its force throughout the play. The outcome can no longer be predicted with certainty, and the game takes on another dimension. In "open" games, a similar effect can be achieved by complicating the rules and expanding the range of possible combinations. In bridge, the fact that some of the cards are laid out on the table is compensated for by a far more complex set of rules than exists in most other card games.

The characteristic features of any game are determined by its rules, rules that create a closed world removed from reality. These rules are usually arbitrary in nature, but the historical and traditional forces underlying them are obvious. In chess and checkers, there are no symbolic births, but the taking of pieces clearly represents some kind of death. These deaths are, just as clearly, not the result of natural causes but of sieges and battles. The course of the game does not reflect any natural evolutionary process. It represents instead a facet of human behavior manifested in a specific historical guise.

We may have created the impression here that the motivating force behind all games is competition or even outright battle. But that is not the case. There are games like solitaire that are played by one person, and there are also games based on cooperation rather than rivalry.

The Japanese royal family used to play a game called "kemari" (*ke* = kick, *mari* = ball), a kind of football in which all the players move around the playing field in a circle and try to kick the ball to each other in such a way that it never touches the ground. Each player concentrates all his energies on not being the one to let the ball drop. We might well wish that this were the game on which the rules of politics were modeled.

Children begin playing purely for the sake of play. Bernhard Hassenstein[6] calls attention to the autotelic quality of children's play, but thinks its teleonomic value lies in the training for life it provides. But Hassenstein does not discuss an essential quality of play: its intensity, its capacity to generate excitement. This is the aspect of play that Johan Huizinga stresses:

> Why does the baby crow with pleasure? . . . Nature, so our reasoning mind tells us, could just as easily have given her children all those useful functions of discharging super-abundant energy, of relaxing after exertion, of training for the demands of life, of compensating for unfulfilled longings, etc., in the form of purely mechanical exercises and reactions. But no, she gave us play, with its tension, its mirth, and its fun. . . .
>
> We play and know that we play, so we must be more than merely rational beings, for play is irrational.[2]

Perhaps we would do better to say that play is beyond reason. It makes sense only in and of itself. As Hamm puts it in Samuel Beckett's *End Game:* "Since that's the way we're playing it . . . let's play it that way. . . ."

3

Microcosm—
Macrocosm

We call the world of elementary particles, atoms, and molecules the microcosm. All the basic physical processes enact themselves in this world. Chance originates in the unpredictability of these processes in their individual manifestations. Only when vast numbers are involved do these processes begin to show predictable patterns and make up what we recognize as the behavior of matter in the macrocosm. Statistical analysis allows us to calculate average behavior from what seems to be the random, unpredictable behavior of individual particles. Viewed statistically, this average behavior is fully predictable, but under certain conditions, the random behavior of individual particles in the microcosm can also affect the macrocosm and introduce there the unpredictability that is characteristic of the ongoing game of chance played at the microcosmic level.

3.1 CHANCE

Elementary processes in the realm of matter are not subject to precise description in terms of time and space. Predictions about the location or velocity of the smallest particles of matter are never any more than statements of probability. This basic postulate of modern physics underlies Werner Heisenberg's formulation of the uncertainty principle in quantum mechanics. This principle states that it is impossible to determine exactly for any given moment both where something is and how fast it is moving. The uncertainty principle had already been noted in classical wave theory. If we want to identify a sound accurately, we have to listen to it for a certain span of time, and if we want to record the sound waves on an oscilloscope, we have to expose the instrument to the sound for one or more complete periods of oscillations. Unless we do this, we cannot accurately determine the frequency, i.e., the number of waves emitted in any given unit of time.

The physicist Richard Feynman[7] has used a paradox of classical physics to illustrate the uncertainty principle. As we all know, positive and negative charges attract each other, and the smaller the distance between them is, the stronger the pull will be. The quantitative formulation of this principle is known as Coulomb's law, named after the French physicist Charles Augustin de Coulomb (1736–1806). Why is it, then, that the positive and negative charges in atoms are so clearly separated from each other? The reason is not that the positively charged nucleus takes up all the space of the atom and therefore prevents the negatively charged electrons from associating more closely with the nucleus. It will help us grasp the relationship between the nucleus of an atom and the diameter of the orbits its electrons travel if we compare the nucleus to the head of a pin centered in the dome of St. Peter's Basilica.

So what is it that prevents the electrons from falling into the nucleus, merging with it, and, at the same time, radiating away their electromagnetic energy? Feynman writes:

The answer has to do with the quantum effects. If we try to confine our electrons in a region that is very close to the protons, then according to the uncertainty principle they must have some mean square momentum which is larger the more we try to confine them. It is this motion, required by the laws of quantum mechanics, that keeps the electrical attraction from bringing the charges any closer together.

A heuristic argument of this kind shows, of course—and this is our point here—only how knowledge of observed phenomena can prevent us from making those untenable abstractions our minds are so given to. Abstractions like point, infinity, continuity, parallelism, and so on are standard concepts in mathematics, but in physics they have to be applied with great caution.

The great epistemological gain that the uncertainty principle brought with it is the insight that we must sometimes abandon the attempt to interpret. Thinking in terms of probability had been part of classical physics since the days of James Clark Maxwell and Ludwig Boltzmann, but it was still generally assumed that statistical uncertainty was not a fact of nature but arose from a lack of detailed knowledge, knowledge that could eventually be uncovered. Innumerable interpretations of the uncertainty principle were offered, all inspired by something like a "quest for hidden parameters." But what finally remained was only the principle itself as it reflected experience.

Interpretation is a process that takes place in our brains, and as a rule it is a cooperative effort. "Science" interprets. That means that a number of minds agree that in a given phenomenon there is something that occurs with regularity, can be reproduced, and can be traced back to recognizable causes, something, indeed, that can be "interpreted."

The brain is primarily an organ for storing information, but the storing that takes place is done according to selective criteria. The brain evaluates the information that comes to it by way of the senses. It filters it, sorts it, compares it with information already on hand, recombines it, and finally assigns it to its place. Karl Popper[8] sees this selective process as inherently deductive and thinks it analogous to the "learning process" of evolution, in which mutations resulting from genetic misreadings are subjected to a constant testing of their viability and, with few exceptions, are proved to be "false" readings.

It is the inner consistency of all natural processes, a consistency of image and mirror image, that makes abstraction possible at all. Without this consistency, events reflecting natural laws could not be reproduced and similarly comprehended by different minds, and their validity agreed upon. A large part of understanding consists in the perception of given phenomena and the recognition, through repeated exposure, of the consistent patterns in them. In the last analysis, we can interpret only what we can experience and what we can subject to experiment.

The insights of quantum mechanics opened up a new epistemological dimension. Shortly before these insights gained general acceptance, Ludwig Wittgenstein made the following dogmatic statement: "Whatever can be said at all can be said clearly, and whatever cannot be said clearly should not be said at all."[9] But what can possibly be said with clarity if any statement we make applies only within certain limits? Even the pioneers of modern physics, scientists like Max Planck, Albert Einstein, and Erwin Schrödinger, the founder of wave mechanics, were never able to make their peace with the ultimate consequences of the uncertainty principle. They refused to believe that nature arrives at her decisions by rolling dice.[10]

In his book *The Logic of Scientific Discovery,* Karl Popper says on the subject of law and chance:

One sometimes hears it said that the movements of the planets obey strict laws, whilst the fall of a die is fortuitous, or subject to chance. In my view the difference lies in the fact that we have so far been able to predict the movement of the planets successfully, but not the individual results of throwing dice.

In order to deduce predictions one needs laws and initial conditions; if no suitable laws are available or if the initial conditions cannot be ascertained, the scientific way of predicting breaks down. In throwing dice, what we lack is, clearly, sufficient knowledge of initial conditions. With sufficiently precise measurements of initial conditions it would be possible to make predictions in this case also; but the rules for correct dicing (shaking the dice-box) are so chosen as to prevent us from measuring initial conditions.[8]

A great many processes in the natural world elude exact description simply because we cannot establish their initial or boundary conditions. But the comparison Popper chose fails to take one factor into account that is essential for predictability and that cannot be ascribed to the particular form of the boundary conditions.

Perhaps we can throw additional light on Popper's comparison by means of another comparison. If we are mountain-climbing, we will behave differently if we are making our approach to the peak through a valley or if we are approaching the summit itself along a narrow ridge. In the valley, there is nothing to worry about. It doesn't matter if we stray from our path; we will automatically be guided back to it. But things look very different if we are on a knife-edge. There, one false step can bring disaster.

The planets are quite stable in their orbits. Minor disturbances will not have catastrophic effects on their courses. But a roll of the dice is made up of innumerable unstable phases.

Using differential topology as a basis, the French mathematician René Thom developed a theory that allows us to analyze the structural stability of physical systems, and he appropriately named it "catastrophe theory." The course that a rolled die takes passes through a number of junctures at any one of which the most minute disturbance will affect the further course of that roll. If we had a die whose edges were so fine that their dimensions were atomic, then the source of any disturbances would have to be sought in the noise spectrum of the thermal motion of the atoms located on the edges of the die. But the uncertainty principle makes it impossible to predict in detail just how those atoms will behave. The best we can do is establish averages. For a stable system, these averages would suffice to predict its macroscopic behavior with certainty. But this is not true for an unstable system. Such a system is affected by microscopic disturbances that can mount up in such a way that the macroscopic level reflects chance activity at the microscopic level.

These differences that originate in the type of phenomenon under consideration will prove to be of the greatest significance for our further explorations of decisions based on chance.

All the direct impressions and experiences we receive through our senses reflect the macrocosm. But even in this realm, we have learned that events that are random if viewed individually are subject to deterministic laws in the context of large numbers.

In the insurance business, it is the customer who assumes almost all the risk, not the company, because the customer cannot know whether the policy will ever benefit him or not. The company will see to it that it always profits, and to guarantee that it always will, it bases its calculations on statistics and applies the law of large numbers. This analogy, like all analogies, is imperfect. It is conceivable that the insurance company will go bankrupt. In the statistical behavior of matter, however, there is a state we call equilibrium in which we can say, with a probability approaching certainty, that "bankruptcy" will not occur. The insurance business does not fall under this category of statistics.

Our considerations up to this point have yielded the following important conclusion: All processes in the physical world—inorganic as well as organic—are based on the motions of atoms and molecules, motions that are, to a certain extent, unpredictable.

But if everything is left to chance, if everything is determined by the roll of the dice, then who or what introduces order? How can a given state recur consistently if the elementary processes behind it can follow any number of possible alternatives? We might expect that the decision-making process, like that in chess, would soon become hopelessly entangled in the innumerable branches of the decision tree.

3.2 THE GAME OF "LIFE" AND "DEATH"

We can begin with a concrete example. The macromolecular structure of protein molecules that have a highly specific function in our organism has to be maintained in every detail. The organism would die without the function that any of these molecules fulfills. But we know that material structures and states have a limited life expectancy. The individual protein molecule has a life expectancy that is very short in comparison to that of the whole organism. Protein molecules are constantly being broken down, and identical molecules therefore have to be constantly built up again. Only in this way can the state of the protein molecule, which is dependent on the molecule's structure and suited to the molecule's function within the organism, be maintained over a long period of time.

Entire organisms and even entire populations have to renew themselves in this same way if they are to survive.

Population density—the number of individuals in a limited space—is a function of the ratio of births to deaths. This is true not only of all living organisms but also of individual cells within an organism and even of molecules and atoms. But on the microcosmic level the physicist or chemist speaks of formation and decomposition, not of birth and death.

At the moment, human births far exceed deaths, and we find ourselves in the midst of a population explosion. But in the animal world, there are sometimes more deaths than births, and a number of species are threatened with extinction.

If a population is to remain constant, there has to be the same number of births and deaths in a given time period. If two sums that consist of a large number of identical units are to be equal, each unit on one side of the ledger must have its equivalent on the other side. But because of the role probability plays in individual physical processes—and, ultimately, because of the uncertainty principle—this kind of balance is impossible. Even with the aid of a built-in control mechanism, we could achieve no more than an average balance over a long period of time. If it were possible to count the number of people in a stable population at various moments, we would see that the actual numbers would sometimes be smaller and sometimes larger than their stable average.

What conditions are necessary if a stable order is to develop in a physical system? This is a basic question that is as significant in the statistical equilibrium of matter as it is in biological or social organization.

As an example, we can imagine a game of "life" and "death" in which the two opponents cast dice to decide the fate of each individual. What strategies can they adopt, and what influence will these strategies have on the populations involved?

The only observable facts on which the players can base their strategies are the population sizes. In theory, there are three possible reactions to changes in population size:

1. *There is an average probability for birth or death that is independent of population size or of changes in population size.* In game theory, we call this way of reacting an "indifferent strategy," symbolized by S_0. The "0" indicates that the birth and death rates are *not* influenced by size of population. An indifferent strategy of this kind is the rule at the molecular level, but in the area of biology, it rarely occurs. Wherever many people live close together, for example, there will be a correspondingly high number of births and deaths.

2. *Birth and death rates are affected by the size of population. A change in population size will bring about a corresponding change in the birth and/or death rate.* If a population grows, then birth and/or death rates also rise. If the population declines, birth and/or death rates also decline. This strategy, symbolized by S_+ is what we call a "conforming strategy," and all it indicates is that the birth and/or death rates will follow the direction of changes in population size. But it says nothing about the proportions involved. They do not necessarily have to stand in a simple (for example, linear) relationship, although that pattern often does occur in nature.

3. *Changes in population size evoke inverse trends in birth and/or death rates.* This pattern is called a "contrary strategy" and is symbolized by S_-. In this pattern, when the population grows, the birth and/or death rates sink. Conversely, the rates rise in a shrinking population.

This inverse reaction amounts to a genuine compensating effect only when we are dealing with the birth rate. Any change in population size will be negated by the contrary change in birth rate. But if this strategy is applied to the death rate, the result is a magnification of the basic trend in the population. These correlations are reversed in the conforming strategy S_+.

We should emphasize here that all these strategies do is define three possible patterns of reaction. In concrete cases, the nature of the species in question will determine which of these strategies is applied.

At the microcosmic level, the choice is made by means of chemical properties or reaction mechanisms that are specific to every class of molecules and that originate in physical forces. The complexity of nature is so great that all these strategies occur in chemical reactions, whether these reactions build up or break down molecules. There are even molecules whose structures and strategies change with their environments. "Intelligent" molecules of this kind are found primarily in biochemistry.

The behavior of primitive organisms is for the most part predetermined, too. Indeed, only by making *selective* use of the molecular strategies offered in nature were they able, in the early stages of evolution, to assume certain key functions.

Within limits, human beings are able to adapt their strategies to the needs of changing situations and to regulate their social orders. In what way can these different strategies affect the size of populations? If we pit "life" and "death" against each other and view their contest from the perspective of game theory, we can set up a pay-off matrix that will enable us to see what possible fates might overtake a population. (See the next page.) This is not, of course, a pay-off matrix in the strict sense of the term. This matrix cannot explicitly define possible gains and losses because this life and death game is not a finite, zero-sum game, nor are its strategies strictly quantitative—that is, the results the possible strategies will bring cannot be predicted in numerical terms. Still, the qualitative insight such a matrix provides can be instructive. From it, we see that there are three possible patterns of population behavior, and we can label them "stable," "indifferent," and "unstable." The fourth category, which we have called "variable," is a composite of the first three. In the "variable" pattern, any one of the other three may, depending on the quantitative context, become dominant and bring about a basic change in the entire system. But what do we mean concretely by each of these terms?

"LIFE"

	S_+	S_0	S_-
S_+	$++$ variable	$+0$ stable	$+-$ stable
S_0	$0+$ unstable	00 indifferent	$0-$ stable
S_-	$-+$ unstable	-0 unstable	$--$ variable

"DEATH"

1. *Stable:* This means that in a population of a given size, birth and death rates will vary to maintain that size. Even the slightest change in the population size, whether growth or decline, will influence either the birth or death rate or both at once in such a way that the change will be compensated for. A "stable" population is thus a self-regulating one. Built-in safety devices protect it from disaster. In populations that are large enough, this self-regulating quality always results in predictable population size.

2. *Indifferent:* The population can assume any size whatever. The average birth and death rates are equal to each other, regardless of the population size. There are no advancing or retarding forces at work. Any size the population takes can function as a starting point for further fluctuation. There is no self-regulation whatsoever.

3. *Unstable:* A population in which the birth and death rates compensate each other becomes "unstable" if a very minor change in population size so affects one or both rates that they accelerate this change. This snowball effect leads to victory for either "life" or "death." In other words, there is a population explosion, or the species dies out. If a combination of strategies labeled "unstable" in the matrix occurs, then a "life" or "death" catastrophe is inevitable, and the fate of the population concerned is unequivocally determined.

These three states appear in the pay-off matrix on the so-called secondary diagonal, represented by the dotted line, but the more interesting possibilities appear on the primary diagonal or solid line. We have already mentioned that the term "variable" describes a state that includes the other three and, depending on quantitative factors, will exhibit the effects of any one of the three. Here the mathematician runs into some non-trivial stability problems. Regulating mechanisms and self-organization in biology and technology are based primarily on the states appearing on the main diagonal. We find these combinations of strategies in the biological macromolecules of self-reproducing structures, where they direct the course of evolution, and we also find them controlling the differentiation of cells and the morphogenesis of organisms. Finally, they determine the organization of information in the network of our nerve cells as well as in computer feedback systems.

4

Statistical Bead Games

Bead games can be used to simulate basic statistical processes. Beads of different colors can represent different kinds of atoms, molecules, organisms, numbers, or letters. A square playing board represents the always limited area within which various processes take place. All the squares in the playing area are identified by coordinates. A pair of dice supplies the element of chance. Depending on the size of the board, the dice will have to be tetrahedrons, cubes, octahedrons, dodecahedrons, or icosahedrons. All squares on the board have an equal chance of being selected by the roll of the dice. The rules of the game channel the chance decisions of the dice and thereby determine the behavior patterns of different populations: indifferent drifting, stable balance, unstable growth, or catastrophic decline.

4.1 "HEADS OR TAILS"

A young man is flying to Europe for the first time in his life. But he has mixed feelings about this trip because he has heard so much about

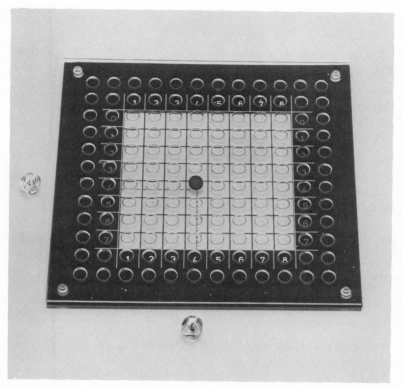

Figure 4. Prototype of a bead game. This board can be used for almost all the games described in this book. The playing surface is divided into 64 squares identified by coordinates. The vertical coordinates are numbered 0 to 7 in red, and the horizontal coordinates are numbered 1 to 8 in black. Two dice in the form of octahedrons are used to determine "chance decisions." The numbers on the surfaces of the dice correspond to coordinates on the board. Every roll of the dice selects a pair of coordinates and, consequently, identifies a bead on the board. The a priori probability of being selected is equal for all squares. In the example shown here, the dice selected the coordinates 4-4, and the blue bead has been placed on the appropriate square.

hijackings and bomb threats. To be on the safe side, he calls his insurance agent and inquires about the risks involved in this flight. The agent assures him that the chances of a bomb being hidden on his plane are quite minimal. But the customer remains uneasy: even if the chance is one in ten thousand, one in a hundred thousand, or even smaller yet, he still thinks the risk is too great. He then asks the agent what the odds are that there will be *two* bombs on his plane. The rather astonished agent replies, correctly, that the answer is the square of the previous figure—if the chance before was one in ten thousand, it is now one in a hundred million. This answer seems to satisfy the young man.

A few weeks later, the insurance agent reads in the paper that a luggage inspection at the airport has turned up a bomb in a passenger's suitcase. The passenger, the article goes on, has claimed in court that he took the bomb with him to reduce the risk of another bomb being on board.

If we use the flip of a coin as our decision-making mechanism in a bead game, we can see what the results of consistently applied indifferent strategies will be.

Table 2. BEAD GAME ''RANDOM WALK''

This game is played by two persons using a square playing board with 16 unnumbered squares on it. Each player has 16 beads, each set of a different color, white and black, for example. Each player places 8 of his beads on his half of the board and keeps his other 8 in reserve. At the beginning of the game, then, half of the board is covered with white beads, half with black. Now a coin is tossed. If "heads" is up, white can remove any black bead and replace it with a white one from his reserves. If the toss is "tails," a black bead replaces a white. The game is played long enough to give both players an equal chance to win.

The game can be further structured and made "finite" by introducing a supplementary rule that gives neighboring beads of the same color an advantage. This rule is that a player may replace on the board any of his opponent's beads that he has completely encircled. The game is over when one color of beads—black, for example—has been completely removed from the board; for in that case, any black bead introduced on the board would automatically be surrounded by white ones. This game shows the physicist how cooperative effects account for the definiteness of the three physical states (solid, liquid, and gaseous) of matter.

"Heads or tails!" Which of us has not at some time or another left a decision to the toss of a coin?

The important thing to keep in mind here is that the result of any toss is completely independent of all the tosses that have gone before it. Even if "heads" has turned up ten times in a row, the chances of "tails" appearing at the next toss remain fifty-fifty. On the playing board, we can analyze quantitatively what the long-term results of a series of independent tosses will be.

In its basic, non-cooperative form, the game "Random Walk" is not particularly entertaining. But if we want to find out for ourselves what statistical consequences result from the consistent use of the indifferent strategy S_0, i.e., a strategy independent of the number and placement of different colored beads on the board, we will need considerable patience. In principle, the game never does end. But if it is broken off too early, the results can be misleading.

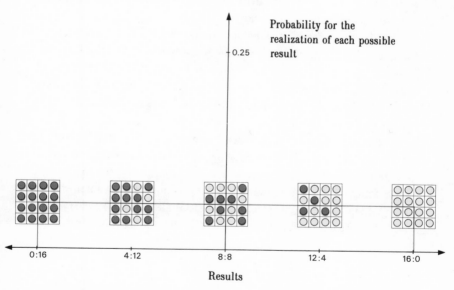

Figure 5. "Random Walk." The game exhibits an uncontrolled process of fluctuation. The probability distribution for the blue and yellow beads is uniform, that is to say, all possible deviations from equal division into eight blue and eight yellow beads are equally probable.

At any point in the game, the probability that either color of beads will increase or decrease is the same. In the early phases of play, the neutral beginning situation of equal numbers of white and black beads will recur with greater frequency than any other. But that does not mean that in the long run the initial situation occurs with greater probability than any asymmetrical division. Any division will show this kind of "short memory," which is to say that any division will tend to recur several times before it gradually dissolves. Because this is true of any possible division, there will be no preference given, after a long period of time, to any particular division. The neutral initial situation will recur only as often or, more accurately stated, as seldom as any other. This is equally true of the "bankruptcy" situation in which one player loses all his beads to the other. This situation by no means ends the game because a player does not have to have any of his beads on the board to be able to continue rolling the dice. The result is that the beads will drift randomly through all possible divisions until the game is arbitrarily broken off.

This drifting process is called a "random walk," hence the name of the game. Mathematically, it produces a probability distribution that has the form of the rectangle shown in Figure 5. This curve shows graphically what we have already expressed logically: *The probability (ordinate) for any possible division (abscissa) is equally great. No division is preferred.*

We can see now what logical error our man with the bomb committed. The probability the insurance agent assumed is analogous to the probability involved in tossing a coin, although the probability of a bomb being on the young man's plane was considerably smaller than fifty percent. No tinkering with the initial situation—by carrying a bomb in one's suitcase, for example—will affect this probability at all.

4.2 THE EHRENFEST URN MODEL

At the beginning of this century, physicists concerned themselves with a less dramatic problem than bombs on airplanes: If two dogs are running side by side and one of them has a lot of fleas, how long will it take until the fleas, by jumping back and forth, have distributed themselves evenly between the two dogs, and how does this equilibrium distribution evolve? It is assumed that the fleas have no particular preference for either dog, although such a preference could be incorporated into the model.

This "dog-flea model"—as scientists sometimes refer to it in jest—arose from statistical studies that the physicists Paul and Tatyana Ehrenfest made in 1907 on how equilibriums are established. The "Ehrenfest Urn Model" is the more serious name under which this statistical game is generally known. For this game, the Ehrenfests distributed a certain number of beads between two urns, but we can survey the game better if we play it on a board with coordinates. (See Table 3 on p. 36.)

The alert reader will soon perceive that in the "non-cooperative" variations of the Ehrenfest urn game the conforming strategy S_+ applies consistently to the "death rate" and the contrary strategy S_- to the "birth rate." In other words, the more one color of bead dominates the board, the greater the chances of "deaths" are for that color; conversely, the likelihood of "births" in that color is smaller. The strategy S_-, like the indifferent strategy S_0 as it appeared in the game of "Random Walk," is atypical for living organisms, and the inherent stability postulated in the pay-off matrix—at least in the form projected by the model—does not occur either in the genetic mechanisms affecting population or in human social behavior. But these strategies are highly relevant for physicists and chemists studying molecular distributions, and this is precisely why Paul and Tatyana Ehrenfest devoted so much attention to problems of this kind.

The course of play in all four versions of the Ehrenfest game concretely exemplifies interesting physical phenomena. Because the moves of the game depend essentially on the distribution of the beads on the board, players are not necessary. All that is needed is someone to roll the dice and move the beads according to the rules. Games of this kind organize themselves, as it were.

Strategic skill is of some use only in the cooperative version, and only in this version does the game come to an end that is predeter-

Table 3. BEAD GAME "EQUILIBRIUM"

For this game, any square playing board with coordinates and appropriate dice can be used, e.g., a board with 16 squares (4 by 4) and two tetrahedral dice, a board with 36 squares (6 by 6) and two dice in cube form, or a board with 64 squares (8 by 8) and two octahedral dice. The beads have to be of at least two colors, and there have to be enough beads of each color to be able to occupy all the squares on the board.

Version 1. Two players take turns placing their beads randomly on the board until all the squares are filled. Then the dice are rolled. The bead on the square selected by the dice is replaced by a bead from the opponent's reserves. The player who has just placed a bead from his reserves on the board also receives one point. Points are recorded after every roll of the dice, and at every roll the "winning" player is the one who has more beads on the board. It is his points that are recorded, and he receives one point for every bead in excess of half the squares on the board. The end of the game is arbitrarily set, and each player then adds his points and divides the sum by the number of times the dice have been rolled. Whoever has the higher score is the winner. The result of the game thus reflects a physical phenomenon. It represents something like an "average absolute fluctuation."

Version 2. The primary purpose of this version of the game is to simulate an approach toward equilibrium. The rules of the game are basically those of Version 1 except that here the game begins with the board completely filled with beads of only one color. The game then proceeds as in Version 1, but now not the excess beads are counted but the *number of rolls of the dice* needed to replace half the original beads with beads of the other color. A roll is counted only for the player introducing a new bead onto the board. A match consists of an even—and preferably large—number of rounds. Whoever ends up with fewest points is the winner.

Version 3. This version resembles Version 1. The difference is, however, that there are four players in this version and four colors of beads. In addition, the game requires a tetrahedral die whose four surfaces correspond to the four colors of the beads, and this die determines at each roll what color bead will replace the bead selected by the numerical dice. This version shows that the results typical for Version 1 apply generally and are not limited to the interplay of two species.

Version 4. In this variation, a "cooperative rule" is added to the basic rules of Version 1: A bead selected by the dice can be replaced on the board only if at least four beads of a different color are adjacent to it. At the beginning of the game, the players take turns placing their beads on the board, but this version differs from the three others in that the placement of the beads has some strategic bearing on the outcome. The "cooperative" effect that neighboring beads have leads to the formation of patterns. We will study this phenomenon in detail in Chapter 6.

---

mined. We will come back to this version of the game later, when we discuss the formation of structures.

Versions 1 and 2 of our game are particularly relevant to the Ehrenfest model. In these versions, we are not concerned with the distribution of the beads on the board. We have identified individual beads on the board only to ensure the same a priori probability for any bead to be exchanged. How they are distributed is of no importance. What interests us is the *excess* of any one color of beads, that is, any deviation from an equal division. In Version 1, we deliberately defined a win in such a way that—if the game is played a number of times and the scores averaged—it reflects a clear-cut result meaningful in terms of physics. It might be useful at this point to take a closer look at a "typical" outcome.

The chances for both (or all) colors to win are always the same. And in Version 1, both (or all) players will, on the average, win or lose with the same frequency. That was true in the "Random Walk" game, too; but, taken over the long run, all possible distributions of beads occurred with equal probability there. This is not the case here. If we use a large enough playing board, one with 64 squares, for example, *the game will always end with an approximately even distribution of both (or all) colors.*

But what do we mean here by "approximately"?

The players' average scores, which we arrive at by dividing the sum of their points by the number of rolls of the dice, reflect directly the average deviation from the initial equal distribution of beads. The typical size of the average score will depend on the total number of beads used in the game. To demonstrate this, we can play the game on boards having 16, 36, 64, and 100 squares. We will find that the average score, seen absolutely, increases as the size of the board increases but not in direct proportion to the number of beads used. Its progression is much slower than a linear one would be. On a board with 64 squares, the average score will be only about twice as large as it is on a board with 16 squares.

The qualifying word "about" reflects the fact that in statistics precise statements can be made only for very large numbers. Sixty-four beads are almost enough to give precise results; with 16 beads, the results will be somewhat erratic.

With a large number of beads, the average deviation—symbolized by the average score—is proportional to the square root of the average number of beads of one color. (In a game with 64 beads and 2 players, for example, this average number will be 64 divided by 2.) This formula represents an important law of statistical physics. The fact that it contains a proportionality factor not exactly equal to 1 is relatively insignificant. Theoretically, a game using 64 beads of two colors would show an average deviation of about ± 10. This number has to be seen as *relative* to the total number of beads of one color. This law of the square root says that the *relative proximity* to an even distribution will increase as the number of beads increases, i.e., the deviations will decrease in inverse proportion to the square root of the beads of one color. In a game with a million squares, the deviations would be large but only about one-thousandth as large as the total number of beads of one color.

As a rule, nature plays with much larger numbers than that. A cubic centimeter of water, for instance, contains more than 10^{22} water molecules (10^{22} equals a figure one followed by 22 zeros, a number almost impossible to express in words). Even the most delicate measuring devices would be unable to register deviations from an even distribution here. For this reason, chemists can use a "deterministic" law of mass action to describe the dynamic equilibrium among reacting molecules.

The law-like nature of this formula for the average deviation becomes clearer still if we ask how often the extreme deviation occurs in which the entire board is covered by beads of only one color. We can say with a probability bordering on certainty that in the non-cooperative version of the game, this deviation will never occur. Even with 64 beads, the probability of such an extreme deviation is only about one in 10^{19}. (By now, we have an idea of just how large such a number is. We arrive at this number by a process similar to that used in the example on pages 9–10. In that example, the final sum was the result of beginning with a penny and doubling progressively every day for a month. Here, the process is the same, but we would have to continue it for 64 days.)

There is a "regulating factor" in this kind of statistics that accounts for the stable behavior. The greater the excess of any one color of beads is, the greater the probability becomes that the dice will select a bead of that color and thereby reduce the excess. And what one player loses, another gains. Rates of increase and decrease will be governed by the distribution on the board at any given moment, and this leads to a stabilization of the equilibrium.

How quickly such an equilibrium comes about is particularly evident in Version 2. Here we begin with the extreme situation of one color occupying the entire board. At first, practically every roll of the dice will reduce the beads of this color and introduce others onto the board. Only when we begin to approach equilibrium does the tempo of this equalization process slow down and turn into a fluctuating pattern like the one we encountered in "Random Walk." The time it takes for a state characterized by a certain distribution of beads to recur is expressed by the statistical frequency with which this state appears. Equal distribution appears with the greatest frequency; with 64 beads, it will, on the average, tend to appear after every ten rolls of the dice. The fluctuation process becomes erratic—comparable to that in "Random Walk"—as we approach equilibrium. Equilibrium will tend to occur more frequently once it has been established, and it will occur less frequently in periods of greater deviation. This drifting process clearly has a short-term memory, too, but at the same time it has a long-term memory that favors equal distribution and is based on the regulating factor that comes into play during wider fluctuations.

Almost on the basis of observation alone, we can project a probability curve. The curve has its maximum at the equilibrium point, and it drops off on both sides to about half this maximum at a distance equal to the average deviation, corresponding to the square root of the total number of beads. The curve then flattens out very quickly toward zero. The result is the bell-shaped curve illustrated in Figure 6 and familiar to all students who have ever been graded "on a curve." This bell-shaped curve is sometimes called the "Gaussian" distribution, after Carl Friedrich Gauss, one of the great mathematical geniuses of all time. If we compare Gauss's bell-shaped curve with the rectangular probability distribution found in "Random Walk," we can easily see the difference between these two models.

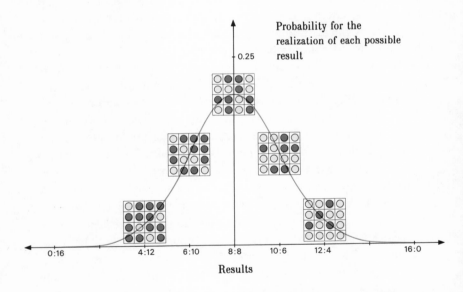

Figure 6. "Equilibrium." This game shows how a stabilizing mechanism resulting from a special combination of strategies produces equilibrium. Gauss's bell-shaped curve gives the likelihood of appearance of blue and yellow beads for any one game. The average deviation from an equal distribution is proportional to the square root of the number of beads of one color when both colors are equally represented.

4.3   THE LAW OF LARGE NUMBERS

Using continuous functions to represent discrete distributions is meaningful, of course, only if we are dealing with large numbers. We can expect predictable behavior only where large numbers are involved. With small numbers, chance determines the results, and both the "Random Walk" and "Equilibrium" models will produce similar patterns. If, for example, two dogs share one flea, only one dog can have it at a time. A truly equal distribution cannot exist at any one moment but only over a period of time. This is the characteristic pattern we found in "Random Walk."

We can now make a summary and comparison of the two models we have studied thus far.

"RANDOM WALK"

1. Both colors of beads appear with equal frequency over a period of time, but there is no preference for an equal distribution, i.e., for both colors having the same number of beads on the board.

2. All the distributions in which beads of one color are in excess appear with equal probability.

3. Since no one distribution of beads is preferred, the equal distribution of both colors has no particular significance. When large numbers of beads are used, it appears relatively infrequently.

"EQUILIBRIUM"

1. Both colors of beads will appear with equal frequency. But here, because of a self-regulating factor that depends on the distribution, an equal distribution is preferred.

2. The average deviation from equal distribution is proportional to the square root of the number of beads of one color when both colors are equally represented. When *large* numbers of beads are used, the relative deviations from an equal distribution will be *small.*

3. Within the range of average deviation, as defined in 2 above, equal distribution occurs with maximum probability and therefore appears more often than any other.

"RANDOM WALK"	"EQUILIBRIUM"
4. The time it takes—or the number of rolls of the dice required—for one distribution to recur is, in the long run, the same for all distributions. There is, however, a "short-term memory" for any distribution that has just occurred.	4. The time it takes for a given distribution to recur is inversely proportional to the probability with which that distribution occurs. Equal distribution will therefore recur most frequently. There is a "long-term memory" for equal distribution.
5. Extreme deviations from equal distribution occur after a finite number of rolls of the dice. This number is, on the average, proportional to the square of the total number of beads.	5. Extreme deviations from equal distribution hardly ever occur even with relatively small numbers of beads. There is a self-regulating factor that works against such deviations, and the greater the deviation is, the more powerful this self-regulating factor becomes.

This summary is based, in every case, on the simplest version of the games involved. The statements in the right-hand column, however, also apply to games with more than two colors of beads (Version 3) and can be adapted to cases in which the a priori probability that beads on certain squares will be replaced differs (Version 4).

This comparison leads us to an important conclusion. The fact that elementary physical events, taken singly, are not determinate by no means invalidates the law of large numbers. It is immaterial whether the unpredictability of single events is inherent in nature or simply a result of our incomplete knowledge.

The matrix on page 28 shows that we can expect stable behavior if either the conforming strategy S_+ on the "death" side or the contrary strategy S_- on the "life" side is set against the indifferent strategy S_0. Stability results if only one process takes over the regulating function, provided the other process does not at the same time counteract this function.

The assumptions we have made for "Equilibrium" are quite realistic. It is obvious that a death rate or rate of decomposition is proportional to the number of given individuals, whether these individuals are atomic nuclei, molecules, cells, or organisms. The more species there are, the more species will die out. This is how the conforming strategy S $_+$ responds to a fluctuation.

How the contrary strategy S $_-$ affects a growth rate is not so obvious. This strategy on the "life" side is characteristic for a state of equilibrium. But true equilibrium can occur only in *closed* systems. We will go into this point in detail in Part II. For now, it will suffice to point out that a closed system is one in which the exchange of energy and matter is zero or strictly controlled. If, within such a system, a certain state A occurs with more than its average frequency —because of some malfunction perhaps—the complementary state B will occur with less than its average frequency. But since A grows out of B, the decline of B will necessarily produce a decline in the growth rate of A. In a closed system, then, the increase of a given species will be at the expense of the material needed for the formation of that species. The contrary strategy S $_-$ is therefore characteristic for formation rates at or near equilibrium, regardless of the kind of reaction or mechanism involved.

In everyday speech, we tend to use the concept of equilibrium, or balance, rather loosely. We speak of an "ecological balance" in the biosphere or of a "balance of power" in politics. The physicist, however, would not use the word "equilibrium" to describe these states, but would instead call them almost-stationary states in which rates of growth and decline maintain a delicate balance but in which genuine reversibility does not occur. The growth of one state does not necessarily take place at the expense of the other. In an open system, an increase can be supported or fed by some outside source, and by-products of disintegration can be channeled out of the system. More often than not, stationary states display the characteristics of equilibrium as defined in our game. In politics, for example, any major violation of the balance of power will evoke retaliatory measures. The same principle applies to a rapidly growing population in a balanced ecosystem. Not only will the death rate increase; but, as a rule, the lack of nourishment will also produce a decline in the birth rate.

Birth control is by no means a human invention, but appears frequently in nature as well. The population of frogs in a pond is regulated by a pheromone that is excreted by recently hatched tadpoles.

This process can be observed if we place tadpoles at different stages of development together in an aquarium. Even if there is an overabundance of food, only the older generation will survive. The pheromone will automatically cut off the development of the younger generations, which, in a natural environment, would never even have hatched.[11] The population size is thus genetically regulated. This kind of biological regulating device is, of course, quite distinct from the built-in mass action mechanisms that regulate numbers in a genuine equilibrium.

4.4 CATASTROPHES

The regulation of quantities in equilibrium depends on a twofold internal mechanism that reacts to fluctuations. The death rate strategy is in conformance with fluctuations, the birth rate strategy contrary to them. If we reverse these strategies, fluctuations will be aggravated rather than reduced, and we will have unstable conditions.

Table 4. BEAD GAME "ONCE AND FOR ALL"

Since we are already familiar with the Ehrenfest model, we can describe this game in just a few words. In this example, we will use a playing board with 64 squares and two octahedral dice. The only rule that differs is the one affecting the exchange of beads. Here, this rule is reversed. Instead of replacing the bead selected by the dice with a bead of an opposing color, we double the selected bead at the expense of the other color. For example, if the dice pick a square with a white bead on it, white can remove any black bead from the board and replace it with a white one from his reserves. Any number of cooperative variations are possible, the effect of which is merely to speed up the already rapid pace of the game.

Our previous examples make the analysis of "Once and for All" simple. The conforming strategy S_+ now governs the birth rate, and at the same time the contrary strategy S_- governs the death rate. Whoever has a lot will soon have more, and whoever has little will

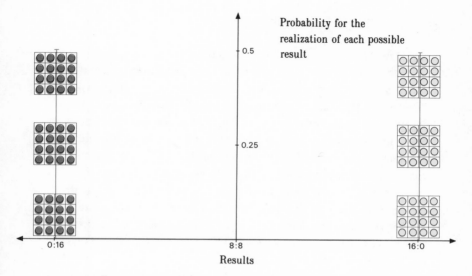

Probability for the realization of each possible result

0.5

0.25

0:16   8:8   16:0

Results

Figure 7. "Once and for All."
Despite equal chances for increase or
decrease for both colors of beads, an
equal distribution in this game is
unstable. Deviations are magnified

and always bring about a "once-and-
for-all" decision. The distribution
at the end of games is not uniform;
instead, either the blue or the yellow
beads dominate the board completely.

lose what little he has (see Figure 7). In the Ehrenfest model, too,
one player gained what the other lost, but the gains and losses were
far less drastic than here.

Both the Ehrenfest model and this inverse variation on it are common in the social realm, but it is not so easy to find examples of the
inverse variation in the behavior of physical particles. In the physical
realm, a conforming strategy S_+ on the life side and a contrary
strategy S_- on the death side are not so likely to occur together, but
even the combination of the conforming strategy S_+ on the life side
and the indifferent strategy S_0 on the death side will suffice to produce
catastrophe.

In the physical world, the conforming strategy S_+ often controls
building processes. The concepts of autocatalysis and reproduction
bring to mind a number of examples. No magic is required to make
two out of one. We can see how one neutron becomes two when a
uranium 235 nucleus absorbs one neutron. The nucleus is split into
two medium-heavy nuclear fragments, and at the same time, two (or

three) neutrons are released. This was the most exciting aspect of the historic discovery that Otto Hahn and Fritz Strassmann made in 1938 and that Lise Meitner and Otto Frisch soon correctly interpreted. What was surprising and unusual in this discovery was that the absorption of a neutron did not make a uranium nucleus heavier or result in the emission of some already known material particle, such as alpha particles. Instead, the nucleus broke into two fragments. The amount of energy released in this process is massive, but that fact would not be of such great significance if a multiplication of neutrons did not take place at the same time. We know from our example of the penny in Chapter 2 what explosive consequences such a doubling process has. And we all know what explosive consequences ensued from Hahn and Strassmann's discovery.

The bead game "Once and for All" can indeed be called a "catastrophic" game. It is over when all the squares are occupied by beads of one color. But this is catastrophe on a minor scale. In the example of the penny, it was crucial that the doubling continue for only thirty days; if it went on for sixty days, it would eat up the currency of even a wealthy industrial nation. If we are thinking in terms of neutrons in a uranium nucleus, what we would call a day in our penny example is equivalent to about a millionth of a second.

In nature, the simple linear form of the conforming strategy S_+ occurs with relative frequency. But there are a number of other more complicated forms of this strategy that lead to similar results. For "Random Walk," we suggested a cooperative variation that can also be used in "Once and for All." Here, too, it will lead to catastrophe, that is, to one color of beads being completely removed from the board. In both games, the players have an equal chance of winning. As long as both players have approximately the same number of beads on the board, it will hardly ever happen that any one bead will be surrounded by four opposing beads, unless, of course, one player had arranged his beads poorly at the start. But once the tide begins to turn in favor of one color—and such a situation can be created by a clever placement of beads—catastrophe for the other color is inevitable.

Physicists copied from nature the regulating mechanism they use in controlled fission. In a nuclear reactor, the consistent doubling strategy of the neutrons is foiled. The splitting of the atomic nucleus releases a huge amount of energy locally, and that energy hurls the resulting neutrons out with unbelievable force. Even if uranium consisted entirely of the fissionable isotope U^{235}, the rapidly escaping neutrons would have to travel quite far before they would be absorbed by another uranium nucleus. In most cases, they would leave the material without having caused any further reaction at all. Only when a "critical mass" has been achieved, i.e., only when it is practically certain that a neutron will collide with another uranium 235 nucleus, can a chain reaction begin and lead to a snowballing of neutrons. This is exactly what happens in an atomic bomb. But atomic reactors use an only slightly enriched uranium in which the fissionable portion (U^{235}) has been raised from 0.7 percent to about 3.0 percent. Here no critical mass is reached, and the neutrons are allowed to escape into the surrounding medium, where they are slowed down by suitable materials, such as heavy water or graphite. After bouncing about for some time, they will finally end their random walk by striking another fissionable nucleus. The "detours" these neutrons take make it possible to regulate their concentration by means of moderators and so to tame the process of fission.

Even non-fissionable uranium 238 atoms can be converted into fissionable material by appropriate reactions in the reactor. So-called fast breeders are able to increase the amount of fuel originally put into them.

The success man has had in taming natural forces might well help solve the dilemma that a very different kind of "catastrophic explosion" presents. As yet, we have not found ways to deal with our population explosion. If we refuse to make use of atomic energy because even our most advanced technology cannot eradicate the element of risk completely, we must consider what catastrophes future generations are likely to face if we fail to solve our energy problem and continue to exploit natural resources.

Nature has her own ways of limiting uncontrolled growth. In doing so, she does not use the combination of strategies we saw in the

Ehrenfest model or in its inverse version. She does not adopt the radically asocial rules of the "Once and for All" model, nor does she go in for the exaggerated socialism of "Random Walk." Instead, she combines strategies to maintain a constant balance between stability, growth, and variability.

5

Darwin and Molecules

Selection is based on a special combination of rules, a combination that allows for variable behavior of whole classes of species, be they molecules or organisms. This behavior includes the random creation of a broad spectrum of variants, the stabilization of an advantageous variant, and the extinction of populations made up of disadvantageous variants. The concept underlying this behavior shows selection, in Darwin's sense of the term, to be a property characteristic of whole categories. Self-reproduction, or reproduction of a complementary or cyclical kind, and metabolism are essential prerequisites. Also, certain "environmental conditions" have to be fulfilled. The rules of the game of evolution are laws of nature, but the historical constraints and the unpredictability of the temporal sequence in which indeterminate elementary events take place produce the uniqueness of all individual detail.

5.1   SELECTION

Let us imagine we are in the railroad station of a large city. Our train will not leave for several hours, and we have plenty of time to visit an old acquaintance. We haven't had any news of him for a long time and have no idea where he is living now. How can we best go about finding this friend?

Surely we would not set out strolling through town hoping to run into him on the street somewhere. The chances of success would be much too small. Nor would we go up and down every street one after the other, looking for his name on mailboxes. With this method, we could be reasonably sure of success, but the investment of time involved would make it impractical. Instead, we would proceed selectively, using certain criteria to determine systematically his possible location.

We may be able to find our friend's address in the telephone book; and if he isn't home, his neighbors will surely know where he works. Before long we will have located him.

The procedure is similar in the game of "Twenty Questions." The players have to identify an object or concept by asking a series of questions that can be answered with yes or no. It is obviously pointless to ask random questions or to use the same set of questions each time. Moving from the general to the ever more specific, the players try with each question to eliminate the largest possible number of alternatives and thus narrow down the possible choices. Each new question will depend on the answer given to the previous one until there are no alternatives left.

In both examples, only one of many conceivable solutions is the correct one, and we want to find this solution as quickly and efficiently as possible. It makes no more sense to simply guess and rely on chance than it does to go through each and every alternative one by one.

In the Ehrenfest model, discussed in the last chapter, all possible states were represented in the average, time-independent probability distribution. But if the number of states is very large, they cannot all occur within a limited time. The chance of finding one specific variant among those that happen to be present is ultimately no greater than in the "Random Walk" model, unless the variant we are looking for occurs with far greater a priori probability than the others. A search procedure based on the principle of equilibrium would be completely inappropriate here, because the alternatives already tested and falsified would, by virtue of their stability, clog the spatially limited system and prevent an evolutionary experimentation with new possibilities.

Living structures are characterized by complexity. The number of conceivable molecular structures for even a single protein is so large that, within the spatial and temporal limits of the entire universe, they cannot even be screened, nor does any one structure occur in large enough numbers to permit the application of equilibrium statistics to it.

If we want to understand how a functional order arises in a system of this complexity, neither the "Equilibrium" nor the "Random Walk" model will be of much use to us. We need instead a model that incorporates instability, for it is instability that forces a clear selection from available alternatives and prevents the return to an earlier, already tested state. This instability cannot, of course, result in an inevitable catastrophe like that in our game of "Once and for All." Instead, the selected variant has to be temporarily stabilized and so provide a basis for testing new alternatives. What we need is a combination of strategies that includes the effects of "Random Walk," "Stabilization," and "Destabilization." Such a combination is present in a game whose three versions lead successively to selection.

Table 5. BEAD GAME "SELECTION"

This game is played on a square board, and the playing squares on the board are identified by coordinates. The game also requires a suitable pair of dice, e.g., two octahedrons. The principle of the game emerges most clearly if a number of different colors, say four or six, are used.

Version 1. All colors of beads are placed on the board in equal numbers at the beginning of the game. They are distributed randomly and fill all the squares on the board. There are enough beads in reserve that any one color could fill the entire board.

Now the dice are rolled, and the following two rules are applied alternately:
1. The bead chosen by the dice is removed from the board and placed in the reservoir.
2. The bead chosen by the dice is doubled, i.e., a bead of the same color is taken from the reservoir and placed on the square that was cleared by the previous roll of the dice.

If this sequence is adhered to strictly, there will always be one empty square on the board after an odd number of rolls, and after an even number the board will be completely filled.

The game is over when one color has filled the entire board. Different values can be assigned to the different colors (e.g., red = 6, blue = 4, green = 2, yellow = 1) so that a winner can be determined if the game is ended prematurely. In other versions of this game, these values are correlated with selective advantages.

Version 2: The rules of Version 1 also apply here, but at every roll of the dice that introduces a reproduction (that is, at every even-numbered roll), a mutation roll is added. Mutation rather than reproduction takes place whenever, in this mutation roll, a specified number, e.g., the "six" on a cubic die, appears. (The remaining numbers continue to represent reproduction.) Thus, if any number from "one" to "five" turns up in the mutation roll, the bead selected is merely doubled. But if "six" turns up, a bead of another color—say the color with the fewest beads on the board—is placed on the empty square. (Other rules can be devised for raising or lowering the rate of mutation.) Assuming that the death rate remains constant, an occasional mutation of this kind will prevent any clear selective trend.

Version 3: The different colors are assigned different values. This means that the average birth and death rates will vary from one color to another. A value die is added to the coordinate dice and rolled with them. For four colors, the following values may be used:

If the bead selected by the coordinate die is:	then beads will be removed and added according to the following numbers turned up on the value die:	
	REMOVAL	DOUBLING
red	1	1, 2, 3, 4, 5, 6
blue	1, 2	1, 2, 3
green	1, 2, 3	1, 2
yellow	1, 2, 3, 4, 5, 6	1

Red has the highest selective value, yellow the lowest. For red the average birth rate is six times as great as the average death rate. For yellow beads, the reverse is true.

Since the assigned values do not permit a strict alternation of removal and doubling, the dice have to be rolled in such a way that the completed processes of removal and doubling do occur alternately. Only in this way can the board remain filled.

The great advantage that red beads have can be counteracted by making the initial distribution unequal. On a board with 64 squares, for example, the game could begin with 2 red, 6 blue, 16 green, and 40 yellow beads. (In nature, too, the occurrence of "valuable" mutants is relatively infrequent.)

In this game, there will always be only one winning color, and it will not necessarily be the one with the highest selective value. If the game is broken off prematurely, the point system for determining the winning color is: red = 6, blue = 4, green = 2, and yellow = 1. Only if the mutation procedure of Version 2 is introduced will red always end up as the winner.

5.2   WHAT DOES "FITTEST" MEAN?

In November 1859, Charles Darwin's *The Origin of Species*,[12] one of the greatest and most controversial works in the literature of science, was published in London. The central idea in this work is the principle of natural selection. In the sixth edition, which appeared in 1872 and which Darwin regarded as the definitive one—the first edition had been prepared hastily because Darwin had learned that Alfred R. Wallace, working independently of him, had come to very similar conclusions—Darwin wrote: "This principle of preservation or the survival of the fittest, I have called Natural Selection."

Unfortunately, Darwin's statement has been frequently misinterpreted, often with dogmatic distortions. Even if we put aside the simplistic reading of the word "fittest" as an absolute measuring stick, applicable even to human beings, we still have to clarify the use of the word "principle."

We cannot use general principles to sum up historical reality any more than we can use the principles of thermodynamics to construct a Rolls-Royce. Darwin did, of course—and this in itself is a major contribution quite apart from his postulation of the principle of natural selection—gather together a vast amount of material about the historical reality of evolution. The principle emerged from that material as an abstraction and was explained more concretely only much later, in particular by the population geneticists John B. S. Haldane, Ronald A. Fisher, and Sewell Wright. Although the so-called neo-Darwinist school was able to formulate clearly in mathematical terms the premises for selection among given populations of species, there is still no agreement even today among biologists about the true nature of the selection principle. Darwin himself began with the concrete fact of living populations. He did not really concern himself with the question of whether the principle was tautological in nature; or, instead, represented an empirical statement that explained the laws inherent in the behavior of living organisms; or, rather, expressed laws deriving from the fundamental properties of matter, laws that are most readily observable in living organisms but are by no means restricted to them. He never, in fact, expressed a clear opinion on this issue, and to which of these views he subscribed is today a matter of merely historical interest. The important point here is that this question can now be answered unequivocally because we now understand the crucial physical and chemical mechanisms underlying life processes.

In cell-free systems that can be maintained under controlled conditions in test tubes, we can reproduce selection and evolution as Darwin defined them.[13] It is a matter of semantics whether we regard the substances that enact these processes, even if they were derived from natural cells, as products of evolution or simply as macromolecules of known chemical composition. Chemists can, in fact, synthesize such structures from the elements. The only reason for using natural substances like enzymes and nucleic acids in our evolution experiments is to save time.

The Darwinian principle of selection can be explained physically, and if conditions and variables are strictly controlled, it can be analyzed with great precision. The bead game "Selection," described in Table 5, demonstrates this.

Version 1 shows that consistent rules of play are more essential to selection than the varying behavior of the individuals or species competing with each other in the process of selection, as provided for in Version 3. We are faced with the paradox that the selective process produces a winner in every game even though competitors do not differ from each other at all. For any game, we can predict the fact that selection will take place, but we cannot predict the specific result, namely, which color will be selected (see Figure 8). The fact that selection takes place follows primarily from applying strategy S_+ to the birth and death rates of the bead populations. Selection is not dependent on varying behavior of the individual beads. But if we were to ask the question "Which of the competitors is the 'fittest'?" we would have to answer: "The one that turns out to be the winner." We have no other criterion for making a judgment than the result of the selection process. In this version of the game, Darwin's principle is reduced to the tautology "survival of the survivor."

If we are to come to a clear understanding of selection, it is essential that we fully grasp this unique example in which we find a negation of the role individual differences usually play in the selection process. This example, which can be observed only if limited space prevents growth, shows that evolution is based on a specific mechanism.

We have, of course, been dealing with an abstract case here that cannot occur in nature in this extreme form. Such behavior would presuppose the complete equality of dynamic properties inherent in

Course of the Game

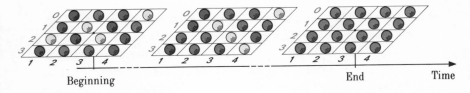

Beginning   End   Time

Figure 8. "Selection." The game
begins with a large number of
different-colored beads that are
represented in approximately equal
numbers. Merely by applying the
strategy S_+ alternately to the birth
and death of the bead selected by the
dice, the game will inevitably lead to
the selection of one color.

all substances and organisms subject to evolution. The tautological
interpretation of the principle of natural selection irritated population
geneticists. In systems as complex as living organisms, it is impossible
to calculate in advance the dynamic factors decisive for birth and
death and so to predict which competitor is "fittest," and it is equally
impossible to define what that concept means in terms of physics. We
are left with the tautology "Whoever or whatever survives is 'fittest.' "

Thus, a precise characterization of selection is possible only under
conditions of extreme competition. The phenomenon of "genetic
drift" shows that these conditions have generally not been met in the
history of evolution. Only in recent years and on the basis of exact
sequential analyses of biological macromolecules in different stages of
phylogenetic development have we been able to understand this phenomenon in all its ramifications. Too narrow a historical view of
Darwin's principle accounts for the fact that this kind of behavior has
been termed "non-Darwinian." Genetic drift is a product of the random behavior of species whose potentials for selection are more or less
equal, and it is understandable in the context of the game "Selection."

Although probability calculations made by the neo-Darwinian
school led to a clarification of the logical structure of Darwin's principle, the reduction of the idea of "fitness" to molecular-kinetic and
thermodynamic parameters and the experimental testing of its relationship to them was first achieved on the level of self-reproducing
macromolecules. These experiments showed that competing individuals may very well differ in physical qualities important for selection,

and these individuals can be assigned a characteristic, physically definable selective value that can be calculated in advance. "Fittest" then indicates an optimal selective value. (We have already defined the term "optimal" on pages 14–15, in connection with von Neumann's game theory.)

Selective behavior, then, will depend primarily on the strategies at work. Birth and death, formation and decomposition, will both make use of the conforming strategy S_+. We have seen that S_+ is the natural strategy for the disintegration or death rate. If every individual has an average life expectancy that is independent of the size of the population, the rate of decline for the whole population will be proportional to the number of existing individuals. In terms of the growth rate, the strategy S_+ results in autocatalytic increase, i.e., reproduction. These are the essential premises for biological evolution. In addition, a permanent energy source, i.e., metabolism, is needed to sustain autocatalytic reproduction. Macromolecules are continuously built up out of "building blocks" rich in energy and broken down into waste products low in energy (see Figure 53). This system must not approach a state of equilibrium, in which, as we have shown with the Ehrenfest model, the contrary strategy S_- controls the death rate. We have elsewhere explained in greater detail these somewhat complex but now completely transparent interrelationships.[14]

Living organisms with metabolism all make use of the growth strategy S_+. Sexual reproduction is only one special type of these widespread autocatalytic mechanisms. On the molecular level, we find other forms of autocatalysis and reproduction. In photography, for example, the copying process involves the intermediary stage of a negative. The molecular carriers of our genetic inheritance, the nucleic acids located in cell nuclei, function similarly. They copy information through the intermediary device of a complementary chain. This process is described in greater detail in Chapter 15. There are also cyclical processes of reproduction. The most interesting current experiments in evolutionary research focus on the "how" of autocatalytic self-organization.

What we should keep in mind is that —even at the molecular level of the organization of matter—there are a number of mechanisms that provide for reproduction or, to put it more generally, for the use of the conforming strategy S_+ for the formation rate. (This is what constitutes inherent autocatalysis.) Because the strategy S_+ can be applied in different ways, such as linear autocatalysis and cyclical or hypercyclical reproduction (see Chapters 11 and 12), the use of S_+ for both the birth and death rates of a population does not by any means always produce the same results. If the birth rate is higher than the death rate, the density of a population can increase at a catastrophic rate. But because a variety of mechanisms come into play, the two rates, given an increasing population density, will change in such a way that the death rate can catch up with the birth rate and stabilize the species or group of species that are best adapted to the environment. In addition, with competitors (or mutants) of comparable viability and therefore of comparable selective value, "Random Walk" patterns of distribution can result. These patterns can give rise to further progress. A condition essential to clear selective decisions is a limited environment or limited amounts of basic materials and sources of energy. In bead games, these factors are represented by the limited size of the board and the limited number of beads.

In nature, too, limits are placed on available space as well as on available amounts of matter and energy, but these boundary conditions are hardly ever so clearly defined as they are in our games. Darwin's principle has been misinterpreted so frequently because the distinction between laws and the conditions under which they operate was not strictly observed. In nature, there is no requirement that a given process stay within defined limits, but evolution will necessarily follow a course that is dictated by a natural law and that can be quantified under specific given boundary conditions. Within this inevitable process, however, the undetermined sequence in which elementary events occur still permits a measure of freedom because the choice of which individual copy will be selected is still open.

In interpreting the results of our games, we have to make use of our imagination. What we can demonstrate in a game with a small number of beads repeats itself in reality with an incredible number of variations. Indeed, it is precisely this complex range of possible physical variations that makes selective mechanisms necessary. Only by means of these mechanisms can those rare viable variants be selected and preserved from extinction. The necessity for these mechanisms exists at the basic level of the biological macromolecules—the nucleic acids and proteins—and underlies the evolution of all living organisms on earth.

Version 1 of "Selection" demonstrates clearly that the conforming strategy S_+ controlling birth and death will result in unequivocal selection. Version 2 shows that, given equal birth and death rates, mutations will introduce an element of variability into the game. Only in Version 3 do we find an evolutionary process that favors a particular species. This process is the product of reproduction, mutation, and selective evaluation, and we could use Darwin's concept of "survival of the fittest" to describe it. "Fittest" here means that selection occurs in accordance with the laws of molecular dynamics and is no longer merely the result of chance, as it was in Version 1. It is still a matter of chance which mutations occur, and in what sequence. But it is a matter of predictable necessity that mutations will occur and that, given mutations, selection will take place among them, defining a gradient of "value." This combination of law and chance suffices to explain the tendency, inherent in evolution, for improvement over time.

In a polemic against the theory of selection we found the following objection: "If someone is intent on reaching a certain destination, he will not take a number of detours on the way [that is, try out random mutations]." It is true that the number of viable mutations is relatively small in comparison to the total number. But it is not true that mutation is purposeful. It is selection that determines the goal of the evolutionary process and thereby avoids detours.

This example clearly shows how dangerous it is to argue by analogy. Analogies can never be substitutes for scientific proof. Once we have arrived at a proof, analogies can be useful in explaining the complex laws the proof has revealed. A proof can only grow out of a careful analysis incorporating demonstrable natural laws. The premises have to be clearly defined, the results subjected to the constant controls of experiment and observation. Incorrect conclusions are, of course, possible, but they will be eliminated once the slightest inconsistency between theory and experience arises.

Is the strategy combination $(++)$ the only one that can explain evolutionary behavior?

Analysis of the pay-off matrix shows that the combinations $(++)$ and $(--)$ produce basically the same results. Either S_+ or S_- could be used, provided the same strategy was applied to both birth and death. But this equivalence exists only *de jure*, not *de facto*. In nature, complex chemical reaction mechanisms are required if the contrary strategy S_- is to be applied simultaneously to both rates. There are enzymes, for example, that can initiate or halt their catalytic functions on signals issued by the substrate. In the reaction cycles of fermentation and respiration in particular, such allosteric enzymes are used extensively to regulate supply and demand. But we are dealing here with unique and highly specialized products of evolution. There is no class of macromolecules that is generally characterized by such a strategy. Furthermore, this kind of regulation can take place only when substrates are present in high concentrations. On the level of very low concentrations—and this is the level of most interest in evolution because every mutant first appears as a single copy—this regulatory method is totally ineffective.

The self-organization of a class of molecules always requires a strategy that is inherent in the entire class. Only in this way is a continuous development possible, and this is why nature practically had to invent the nucleic acids. All living organisms reflect the basic reproductive strategy S_+ of the nucleic acids, and it was this strategy that made the molecular self-organization of living structures possible. Only where regulatory mechanisms programmed in advance occur do we find the contrary strategy S_- in effect for both processes at once. This is true also for certain regulatory elements in man-made machines where this strategy is applied quite frequently.

Darwin, of course, did not have detailed information of this kind at his disposal. With the words "Darwin" and "molecules," the title of this chapter brings together two worlds that could only be linked in our time through work in molecular biology or, more exactly, through the spectacular cooperative efforts of biologists, chemists, and physicists. The hypothesis that Darwin derived from meticulous observation and expressed in a simple formula has been confirmed by natural laws revealed by statistical physics. Both disciplines arose in the same historical era. Ludwig Boltzmann and Charles Darwin were contemporaries. Boltzmann introduced statistics into physics, and Darwin discovered new laws governing the development of living organisms. In Part II of this book, we will come back to the synthesis of these scientific advances and to the physical foundation of the principle of selection.

5.3   THE GAME OF "SURVIVAL"

We have now explored all the basic strategies involved in games. In more or less pure forms, these strategies appear in classic parlor games. These strategies are, of course, subject to special quantitative variations. Cooperative versions of games provide examples of such variations, and in later chapters we will come upon a number of these examples. In concluding this chapter, we will consider, as a prototype of a parlor game, still another selection game in which the strategies discussed so far are combined. The nature of this game prompts us to call it "Survival."

Table 6. BEAD GAME "SURVIVAL"

This game is played on a board with 64 squares, and, as in "Selection," the fate of the beads is determined by rolling dice. Birth, death, competition, and the securing of territory are elements in this game, which proceeds according to the rules discussed in preceding chapters but at the same time permits the player enough freedom to turn the chance decisions of the dice to his own advantage through skillful play. Whoever has the most beads in "secure" positions at the end of the game is the winner.

To begin the game, both players alternately place their beads on the board. Tactics play a role even in this initial phase of the game because the players try to place their beads in a way that will enable them to win as many survival positions as possible. After half the squares (32) on the board are filled, the players begin to roll the dice, again in strict alternation. Now the following rules go into effect:

1. If a player rolls the coordinates of an empty square, he may fill it with a bead from his reserves. If he has no beads left in reserve, he can take one of his disadvantageously placed beads and move it into the square the dice have selected. (If all his beads are already in favorable positions, he may pass.)
2. If the dice select a square occupied by a bead of the opposing color, this bead is removed from the board and placed in the reservoir. But if the bead is in a survival position, it is not removed.
3. If the dice select a square occupied by a player's own bead—regardless of whether it is in a survival position or not—this bead is doubled. The player takes a bead either from his reservoir or from a disadvantageous position and places it on any empty square on the board. In addition, he may roll the dice once more, and the rules spelled out here apply. There is no limit on the number of times a lucky player can roll the dice.

The squares on the board have strategical significance in this game. At the outset of the game, all the squares have equal value, but as the beads are placed, some squares become more advantageous than others. This effect simulates specific interactions between particles of matter.

a. Survival Positions. If four beads of the same color are placed on adjacent squares to form a quadrangular configuration (Figure 9), they are "stable" and cannot die out unless they are surrounded by opposing beads and thus made unstable. These quadrangular survival regions can easily be extended. A new block is formed, for example, if two beads are added to an existing square. Every time a player completes a new block of four, he may remove from the board any one non-stable bead belonging to his opponent and place it in the reservoir. As Figure 9 shows, it is possible to create several blocks of four with one move.

b. Enclosures. If a player can manage to surround any of his opponent's beads, he may remove them from the board and place them in the reservoir. To surround opposing beads, the enclosure formed must consist of a continuous chain of adjacent squares in orthogonal formations (Figure 9), and squares on the edge of the board can also be surrounded.

The game is over when one player has placed all his beads on the board in stable ("survival") positions. All beads in survival regions count as points, and whoever has the highest number of points is the winner.

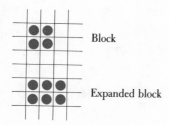

Block

Expanded block

Block completion   Enclosure patterns

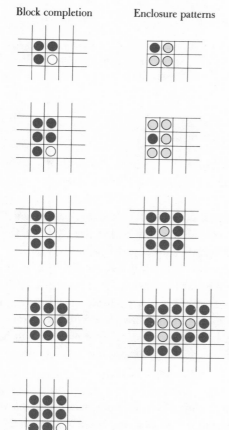

Figure 9. Illustration of various survival positions arising from block completions and some examples of possible enclosure patterns.

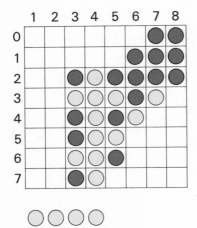

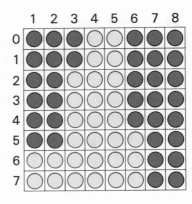

Figure 10. "Survival." The board on the left shows a typical distribution after the initial placing of the beads. Blue has managed to form three overlapping survival blocks, and, as a consequence, yellow has lost four beads. The board on the right shows the distribution at the end of the game. Blue has won by a slim margin. As a rule, the game ends with a less symmetrical formation.

Superficially, this game resembles the Japanese game called go. This is particularly true of the initial phase, in which the beads are placed according to strategic considerations. But all this phase does is create more or less advantageous initial positions for the players. The actual process of selection begins with the rolling of the dice, and at this point the rules allow for all possible statistical fluctuations.

First of all, there is the rule that empty squares selected by the dice can be filled by a player's own beads and that squares occupied by the opponent's beads may be cleared. This rule by itself would result in stable equal distributions like the ones we saw in equilibrium games. If a large number of squares is occupied, then a large number will be selected to be cleared. Filling a previously empty square is equivalent to a "primal creation." But as the board fills up, these creations become less and less significant. The law of reproduction for living structures is autocatalytic in nature. That is, only where organisms already exist can more come into being. The doubling rule in this game reflects this law.

If growth and decline were exactly proportional to the number of occupied squares, population density would fluctuate in an uncontrolled fashion, as it does in the "Random Walk" game. For this reason, we introduce distribution-related "selective advantages": survival regions, repeated rolls of the dice, enclosures, etc. These advantages produce a "non-linear" acceleration of the game.

As in natural selection, the factors of stabilization, drift, and destabilization control the course of the game. The situation resulting from chance decisions determines which factor ultimately dominates in any given game. "Survival" often ends in a near draw. But catastrophes can occur, and it can happen that all of one player's beads die out. Catastrophes of this kind can, to some extent, be prevented or provoked by the players' strategies.

In most parlor games, the players can influence the course of play. The games, however, that we have previously discussed and that are described in Tables 2 through 5, conceived as they are to simulate natural processes, practically "play themselves."

TWO
Games in Time and Space

There has to be a unifying force at work between form and matter, that is, there has to be an impulse of play, because only the oneness of reality and form, of chance and necessity, of suffering and freedom does full justice to the concept of man.

Friedrich von Schiller, "On the Aesthetic Education of Mankind," Letter 15.

6

Structure, Pattern, Shape

The form in which reality presents itself to us is highly structured. Conservative forces fix the products of chance and create permanent forms and patterns. Dynamic states of order emerge from a synchronization of physical and chemical processes, accompanied by the constant dissipation of energy. The living order is based on the "conservative" as well as on the "dissipative" principle. The shape of organisms and the gestalt of ideas both originate in the interplay of chance and law.

Creation began with the emergence of form. "And the earth was without form, and void; and darkness was upon the face of the deep." "The deep"—*t'hom* in Hebrew—represents unformed primal matter, which is often called *mayim,* water, as well. The Bible is not concerned with the actual composition of the primal medium here, but only with its spatial and temporal formlessness. From this formlessness emerge the structures of time and space: light and darkness, day and night, heaven and earth, land and sea.

Form is a product of order in time and space, but it can also manifest itself simply by dividing things into different classes. Something is perceived as a unified shape only when it is more than the sum of its parts. This is a basic characteristic of shape or gestalt, one that Aristotle noted and that his disciples and followers on up to the Neoplatonists repeatedly commented on.

> When all the strings of a lyre are tuned in the Lydian mode and they are plucked individually and so do not sound in harmony, then each one will no doubt produce the correct tone. But the harmony that results from the simultaneous sounding of all the strings is obviously different from the sounds produced by the individual strings. The union of all the strings creates a form that is not present in the broken chord. In other words, the totality that emerges from the harmony of all the strings plucked together (even if they are separated spatially) differs from the totality produced when the strings are plucked individually. But these totalities are the same in the sense that no new note is added when the unison of the strings gives rise to the harmonic form [J. Philoponos, 6th century A.D., cited by S. Sambursky[15]].

The essence of our concept of form or gestalt lies both in its quality of being more than the sum of its parts and in its transmittability. Our brains always interpret a form as an integrated whole. The lenses of our eyes transmit the contours of a spatial structure in their original proportions to the retina. There, these contours are transformed into electrical signals that pass this information on to the brain by means of temporally correlated nerve impulses. The random emission of individual impulses is transformed into an ordered sequence of signals. It is only in the cortex that a spatial pattern of electrical activity again takes shape and is classified and interpreted in terms of the experience stored in the thresholds for excitation of nerve cells.

In a similar way, the basilar membrane of the ear transmits a pattern of sound that is then transformed into a temporally correlated sequence of impulses and conducted by way of separate channels to the perception center, where it is recorded in a spatially fixed engram. We perceive shapes in totalities only by means of correlations that modify the constantly present statistical discharges of the nerve cells in a reproducible form. Thus, a shape is anything that stands out from the statistically uncorrelated flood of stimuli that originate in the space-time world we are able to perceive.

The interplay between the individual parts of an object we perceive as a shape can be highly complex. Indeed, it may be that the perception of something being "more than the sum of its parts" originates in our brain (cf. Chapter 16).

6.1   CONSERVATIVE   STRUCTURES

Spatial structures of matter are always the result of physical forces at work among the parts. These forces alone are responsible for creating a state of order. We have already encountered shape as the result of interacting forces in the bead games of Chapters 4 and 5, and we will use these games as our point of departure here. In the Ehrenfest model, we saw how two sets of beads changed places with each other reversibly on the basis of a statistical mechanism. There is a given a priori probability for each bead to be exchanged. For all the beads taken together, however, the law of mass action brings about an order based on equal distribution. This order is purely numerical and without spatial manifestations. The beads of different colors, although controlled in number to a certain extent, distribute themselves randomly on all the squares of the board. Clusters of one kind of beads are products of pure chance and occur in a non-reproducible manner. A spatially ordered pattern will arise only if we introduce a rule that affects the interplay between neighboring beads and hence influences the a priori probability of single exchanges taking place. Versions of the game that demonstrate cooperative effects are based on rules of this kind. We can, for example—making use of a theory of cooperative transformation worked out by Bragg and Williams—introduce the rule that a black bead selected by the dice can be replaced by a white one (or vice versa) if all eight adjacent squares are occupied by white beads. If that is not the case, the dice are rolled again to decide whether the selected bead can be replaced or not. The condition governing this roll is that for every adjacent different-colored bead or empty square the probability that the selected black bead can be replaced is reduced by one-eighth. A black bead surrounded by all black beads cannot be replaced by a white one (or by one of any other color). The degree of cooperative interplay may, of course, vary. The cooperative effect is weak when only *one* neighboring square is filled with a bead of the same color as the bead selected by the dice. This is the predominating situation at the beginning of play; and in this early phase, the distribution of beads differs hardly at all from that in non-cooperative versions of this game. Only as the interplay between neighboring squares increases do blocks of like-colored beads take shape and expand until one part of the board is occupied only by white beads, the other only by black. In other statistical bead games in which equal distributions do not become stable, as they do in the Ehrenfest model, one color of beads can even take over the entire board. Both the condensation of vapor to liquid and the crystallization that takes place in smelting occur by processes analogous to the one we have just observed in games based on cooperative versions of the Ehrenfest model. With the aid of the bead games we have discussed, the physical theories that underlie these processes—theories like the ones developed by the Russian physicist Lev Landau[16] and his school—can be demonstrated in concrete form.

The precisely defined distribution of atoms in a molecule, the spatial structure of a protein, the symmetrical pattern of crystals, the bizarre shape of a massif, and the pattern of a constellation in the night sky *all result from static forces at work among physical particles* that lend more or less symmetrical forms to the totalities in which they occur. Structures of this type are called "conservative." We speak of a conservative force if we can assign a time-independent potential energy to every point in the force field. If a movement (an oscillation or rotation) should take place in the force field, the sum of kinetic and potential energy will still remain constant.

Conservative structures can be highly complex. A protein molecule provides a striking example. The precisely defined spatial structure of all the parts in this molecule (Figure 11) can be attributed entirely to conservative forces, and we can readily see from this model that conservative structures are not necessarily associated with simple symmetrical operations. Even viruses, which cannot be classified as

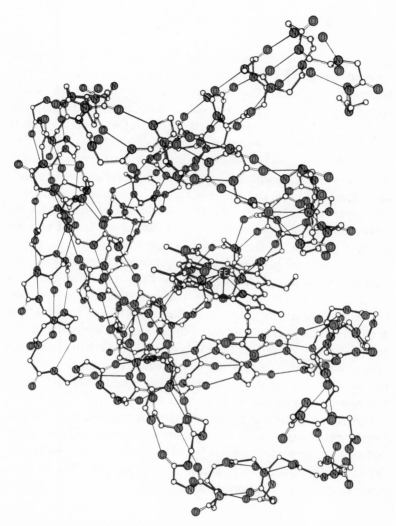

Figure 11. The structure of a protein molecule results from the action of conservative forces that determine the spatial arrangement of every atom. It can be reconstructed with the help of X-ray diffraction patterns. The illustration shows the chain of a hemoglobin molecule in its three-dimensional folds. The active center (shown in the middle of the illustration) takes the form of a "pocket" and surrounds the oxygen molecule attached to the iron complex (Fe). The detailed knowledge we now have of the structure of the hemoglobin molecule is primarily due to the work of Max Perutz,[17] who kindly granted us permission to use this illustration.

either living or non-living but exist somewhere on the borderline between the two, originate their shapes through a complex interplay of conservative forces. The molecular unit of a virus particle can be completely broken down into its individual components (Figure 12), and those components—like the parts of a puzzle that, in nature, puts itself together—can then be reconstructed into an infectious unit. As physical particles, viruses can, like mineral substances, form crystals. But placed in an environment of living cells, they behave like living organisms. They reproduce and multiply, recklessly exploiting the metabolism of the host cell, usually until that cell is destroyed. When we move on to the subject of symmetry in the next chapter, we will have to discuss in more detail the fact that protein molecules and viruses, despite their extremely complicated structures, can indeed be crystallized. We need only study one of Maurits Cornelius Escher's brilliant drawings (Figure 13) to gain an intuitive understanding of this process.

Conservative structures are mostly structures that correspond to an equilibrium; they are characterized by an absolute minimum of free energy. Some, however, are fixed only by a local or relative minimum of energy. Unlike all forms of life, which require a metabolism in order to exist, these conservative structures do not need to dissipate energy to maintain themselves.

Goethe was acutely aware that there were two fundamentally different types of spatial order.[19] In his lectures on comparative anatomy and zoology, he made the following remarks under the heading "On the Laws of Organization as Such, to the Extent That We Can Observe Them in the Structure of Types":

To facilitate our comprehension of the concept of organic existence, let us first take a look at mineral structures. Minerals, whose varied components are so solid and unchanging, do not seem to hold to any limits or order when they combine, although laws do determine these combinations. Different components can be easily separated and recombined into new combinations. These combinations can again be taken apart, and the mineral we thought destroyed can soon be restored to its original perfection.

The main characteristic of minerals that concerns us here is the indifference their components show toward the form of their combination, that

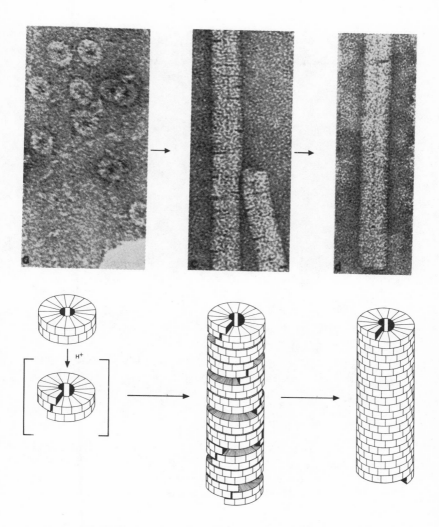

Figure 12. The jigsaw-puzzle process of reconstructing a virus particle can be made visible with the help of the electron microscope. The scale in the photographs in the upper part of this illustration is about 500,000:1. In a neutral solution, the virus protein is present in the form of disc-shaped aggregates (above left). As the solution is made more acid, these aggregates draw together and form small helices at first, then, after about fifteen minutes (above center), larger ones that are still marked by visible "imperfections." After about eighteen hours (above right), all the joints in the structure are perfect. The lower half of the illustration shows this process in schematic form. Aaron Klug conducted this experiment and interpreted its results.[18]

Figure 13. "Development II" (1939) by Maurits Cornelius Escher (1898–1972) (*Die Welten des M. C. Escher* [*The Worlds of M. C. Escher*], Heinz Moos Verlag, Munich). This illustration is used with permission of the "Escher Foundation," Haags Gemeentemuseum, Den Haag.

is, their coordination or subordination. There are, by nature, stronger or weaker bonds between these components, and when they evidence themselves, they resemble attractions between human beings. This is why chemists speak of elective affinities, even though the forces that move mineral components one way or another and create mineral structures are often purely external in origin, which by no means implies that we deny them the delicate portion of nature's vital inspiration that is their due.

How different even imperfect organic beings are! They convert part of the nourishment they absorb—eliminating what they do not need—into distinct organs. What they do absorb they turn into something unique and exquisite by joining most intimately one element with another and so forming differentiated parts in whose forms multifarious life is manifested. And if these forms are destroyed, they cannot be reconstructed from what remains.

If we compare these imperfect organic beings with higher ones, we find that the former, even though they make use of elemental influences with a certain degree of force and individuality, cannot bring the resulting organic parts to the same level of specialization and permanence as the higher animal forms can. We know, for example, that plants—and we will not descend any lower on the scale of organic life—developing as they do in a certain sequence, represent one and the same basic organ in highly different shapes.

Detailed insight into the laws governing this metamorphosis will surely advance the science of botany, not only in its descriptive tasks but also in its effort to understand the inner nature of plants.

(In offering this long quotation from Goethe, we mean to illustrate only his talents as an *observer* of nature. What is so admirable in this quotation is how clearly he distinguishes between the two principles underlying structure in nature. Goethe was not always so acute as an *interpreter* of scientific data.)

It is understandable enough that Goethe was more fascinated by the mutability of organic structures than he was by their invariance or persistence. After all, the goal of his pansynoptic speculations was the constant metamorphosis of human thought and ideas.

Every plant unto thee proclaimeth the laws everlasting,
Every floweret speaks louder and louder to thee;
But if thou here canst decipher the mystic words of the goddess,
Everywhere will they be seen, even though the features are changed
Creeping insects may linger, the eager butterfly hasten,—
Plastic and forming, may man change e'en the figure decreed.

6.2   MORPHOGENESIS

For the working molecular biologist today, the identical reproduction of an organism is no doubt a greater miracle than the occasional metamorphosis of one.

As we have already mentioned, conservative structures, particularly at the subcellular, molecular level, play a major role in morphogenesis. The molecular make-up of the entire information storage system, the double helix, the spatial placement of functional groups in biocatalysts, such as enzymes, indeed, the entire structures of organelles and cells make use of conservative forces that precisely determine the location of every detail. The invariance of the genetic program is a concrete expression of these rigid conservative forces. And occasional variations, or mutations, are nothing more than "misreadings" that occur as the result of thermal fluctuations on this level. These misreadings are preserved by means of identical reproduction, and if they prove to be advantageous, they will be selected. This is the fundamental process that makes evolution possible.

But how then are we to explain the ontogenetic development of an organism that begins its life as a single fertilized egg cell and will grow into a mature individual made up of billions of somatic cells with different specific functions? Is this development comparable to that complex jigsaw-puzzle process by which virus particles assemble themselves (cf. Figure 12)? If this were the case, the information for each of these billions of different cells and their constituent parts would have to be separately programmed genetically. That would be extremely inefficient.

Nature has in fact chosen a different and much more efficient method. Every cell has to have a characteristic sign by which it can be recognized, and the interaction between cells that is set in motion by these signs is purely conservative in nature. But how can individual cells be equipped with characteristic signs if all the cells of the body stem, by way of reproduction, from a single egg cell? The system itself chemically creates a pattern of morphogenes, which are chemical probes of sorts and which, at different stages in the organism's development, draw partial information from one and the same storehouse.

We will have to take a closer look at these fundamental problems of morphology. But first we must look into the question of characteristic signs and determine whether we are merely dealing with an ad hoc hypothesis here.

We are, of course, dealing with a hypothesis, but there is considerable supporting evidence for it. It has been proved that many cells do have such characteristic signs in the form of specialized receptors. How these receptors work has been studied most thoroughly perhaps in immune systems.[20] As Karl Landsteiner recognized as early as the turn of this century, the receptors here are represented by the specific antibodies produced by each cell. Our organism manufactures a vast repertoire of such antibodies. This repertoire is so vast that practically all molecular patterns that occur in nature are represented by a complementary template in the immune system. An invader in the form of an antigen (e.g., an incompatible protein molecule or a bacterium) will not stand a very good chance of evading this "police force" of antibodies. It will almost always be recognized by one or more of its complementary antibody templates, and it will then be made harmless as it is chemically bound and broken down. The relatively few "policemen" on duty, of course, will not be able to handle a massive invasion. In this case, the population of the cells that produce the necessary antibodies has to be increased. This is the process that is triggered by the receptors located on the surfaces of the cells producing antibodies. By binding the invading antigen, the receptors pass on a signal to the synthesizing machinery of their cells which then increases antibody production as well as cell division and thereby initiates a multiplication of the cell type in question. The lymphatic system is soon flooded with a huge number of specialized antibodies that cut short the spread of any bacterial invader. If this whole process functions properly, we say that the organism has become immune.

The immune response is set in motion, then, by the multiplication of a few "scout" cells that are always present. The immune memory consists of a highly complex network of multifarious interactions (cf. pp. 292 ff.).

We could account for the differentiation in organisms by assuming that in each cell—at the same time that the program for metabolism and reproduction common to all cells is selected—a specific function is picked out from a universal program and, by means of a specialized receptor, stamped on the surface of the cell. There are a number of arguments that support this hypothesis:

- Chromosomes, even in human beings, have a "mere" few billion code signals, of which only a relatively small percentage are utilized for coding functions. The programming capacity is nowhere near large enough to allow the genome of the fertilized egg cell to pre-program every single one of the body's billions of somatic cells. This suggests that the somatic cells must be produced according to some hierarchical structural principle.
- In the process of cell division that begins immediately after the egg cell is fertilized, the *entire* genetic program is reproduced in every division. Only later does a specializing process determine which part of the program is to be activated and which left inactive in the development of particular types of cells.
- From the previous point it follows that every somatic cell is equipped with the complete program of the fertilized egg cell, and we can in fact "artificially" reproduce a complete organism by transplanting the nucleus of a somatic cell into an egg cell from which the nucleus has been removed (cf. p. 180). We can also artificially arrest the specializing process in the laboratory.

Impressive experiments in cell determination and differentiation have been conducted at the Max Planck Institute in Tübingen with hydrozoans and slime molds. Günther Gerisch was able to prove the existence of a chemical synchronizing mechanism in the aggregation of amoebae. Alfred Gierer[21] and Hans Meinhardt studied the process of cell determination in the freshwater polyp hydra. This organism, which feeds on bacteria, consists of about one hundred thousand cells. It can be separated into its individual cells without suffering any harm, and these cells can, under the proper conditions, grow again into the original organism at any time. In this experiment, the natural process of pattern formation was simulated, and at the same time new functions could be assigned to cells by means of external manipulations (see Figure 14, page 82). What a cell becomes is not necessarily determined directly in the genetic program. Every cell has the potential to assume practically any function. The information determining specialization must therefore come from the environment of the cell in question. But where does the "environment" get this information? Clearly this information must originate in some cell of the organism. This means that the environment has been structured by a process that this cell has directed. There has to be something like a pattern of concentration of some chemical substance that stimulates the production of specialized morphogenes. The distribution of this concentration thus anticipates certain polarities of the form to be; this distribution is already a form or gestalt itself.

How can a stable spatial pattern arise in a homogeneous fluid medium that has no fixed physical determinants?

To understand better this fundamental problem of morphogenesis, we will have to make a digression into the "dynamics of chemical reactions." Once again, games will help us understand complex interactions that only the language of mathematics can convey with the accuracy that science requires.

6.3 REACTION GAMES

We have already encountered some examples of chemical reactions in earlier chapters. Almost all the bead games we have discussed so far can be easily interpreted as reaction models, and, conversely, we could invent a bead game illustrating any possible chemical reaction.

We will take the Ehrenfest game described on page 36 as our point of departure. It symbolizes the phenomenon of chemical equilibrium as that phenomenon is generally expressed in the law of mass action. The points that originally interested us in this model were the mass ratios that evolved, their average values, and their statistical variances. The reaction between two or more components that concerns us here is obviously a localized process, too, one that depends on the spatial distribution of the reactants. In the cooperative versions of our game, this spatial factor is directly reflected in the formation of patterns on the playing board. What can we say about the spatial occurrence of patterns of this kind that appear only sporadically even in cooperative games? The place of the reaction will depend solely on the presence of a reaction partner and is therefore not absolutely predetermined in space.

But we need to study this process in more detail, and we will focus on two reaction partners we will call "he" and "she." They have to

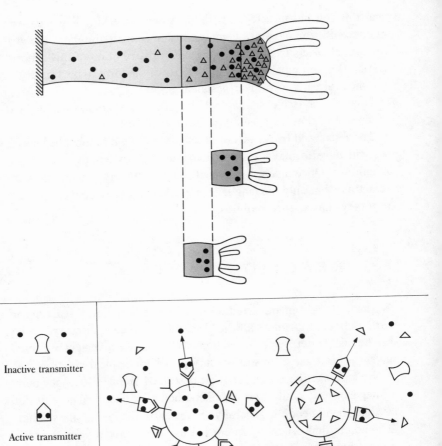

Inactive transmitter

Active transmitter

Y   Receptor of cells that produce the activators
⊤   Receptor of cells that produce the inhibitors

Figure 14. The freshwater polyp hydra is well named. Just as the nine-headed Hydra of Greek mythology grew two new heads for each one that was cut off, the polyp hydra can, if cut into pieces, regenerate itself completely from each of these fragments. Although the cohesion of cellular structures is maintained by conservative forces, it is a dissipative structure that provides the "polarization" necessary for cell differentiation and morphogenesis.

The illustration on the left shows in schematic form an experiment conducted by Alfred Gierer and Hans Meinhardt. A hydra polyp, which has tentacles at the head end, was cut into several sections. The fact that each section grew both a head and a tail from immediately adjacent cell regions contained in that section shows that specialization is caused neither by absolute spatial position nor by the absolute value of concentrations of any particular morphogenes but solely by a gradient in cells or cell components that determines polarity.

The diagram above illustrates a model Gierer and Meinhardt developed to explain their findings. They assumed that there is a gradient of concentration that includes two substances, an activator ($\bullet$) and an inhibitor ($\triangle$). Both substances are produced by different types of cells equipped with receptors. If the inhibitor is present in a high enough concentration, it can block the function of the activator cells by being bound to the receptors of those cells. But if the activator is present in excess, it can stimulate the production of additional activator *and* inhibitor molecules by means of a protein that interacts with both receptor types. In this way, the concentrations are regulated autocatalytically and "competitively." Such a reaction mechanism, which incorporates a non-linear autocatalysis, yields dissipative structures (see Figure 21). These structures determine polarities and thereby guide the differentiation of cells.

meet if a reaction is to take place. They are by no means drawn to each other over great distances, nor does "he" blushingly follow "her" at a distance. They take no notice of each other, not, at any rate, unless they both exert some physical force on each other. But all forces involved in chemical reactions have only very short ranges. As a result, both partners wander aimlessly, now approaching each other, now moving apart. They derive the energy for these random movements from collisions with other particles in thermal motion in the medium, and they are shoved about erratically by those collisions. In time, each partner will cover a certain area, the extent of which—as Albert Einstein showed with the help of statistical calculations—increases in proportion to the square root of the time period involved. Eventually the spheres of influence of the two partners will overlap. In concrete terms, this means that each partner will have been pushed about perhaps a million to a billion times by the molecules of the medium. But that may take only something like a millionth to a thousandth of a second. Hence, it is not long at all before the great moment arrives: The reaction can take place. Given the right temperaments, there is "love at first sight," even among molecules. "Phlegmatic types" first have to pass an activation threshold. They let themselves be moved around a few more times before a successful approach can occur.

The reaction itself can consist of an exchange of electrons, of protons, or of whole parts of the molecules. Sometimes it is no more than a transfer of energy that aids in overcoming an inner barrier. The partners are always somehow changed by the encounter, and at least one of them changes "his/her" name. Some decide to stay together forever. Others separate after a short time; they tire of their new status and simply revert to their old one. It can also happen that several particles emerge from the meeting of two.

But as soon as the partners leave the area in which the forces they exert on each other are effective, they become indifferent to each other. "Out of sight, out of mind." Once again they are moved about randomly by thermal collisions until the game begins anew. There is no personal loyalty among molecules.

Games like Parcheesi, or its considerably more complex predecessor, backgammon, are essentially reaction games. The "random walk" process is maintained by rolling dice. Reactions following on encounters are, of course, less pleasant than in our molecule episode. Here, the whole point is to eliminate one's opponent's pieces from the board.

Backgammon is based on a strategic concept. The game as we know it today is played according to rules established in England in the eighteenth century, when it enjoyed a great resurgence. It originated in ancient times; board games that are forerunners of backgammon were found in the tomb of Tutankhamen (ca. 1350 B.C.). We should

Figure 15. The photographs show three board games and their respective playing pieces, all of ivory, found in the tomb of King Tutankhamen (ca. 1350 B.C.). "Each board is divided into thirty equal squares arranged in three rows of ten squares each. Each game has ten pieces, colored black and white and comparable to the pawns in chess. Each player has five pieces of the same color. Instead of dice, small bones or black and white rodlets were rolled to determine moves. This is an old game of chance obviously related to the game of *El-tab el-siga* (*seegà* = playing board, *táb* = playing sticks), which is popular throughout the Orient today. From the rules of the modern game, we can deduce how this ancient form of the game was played. It required little or no special skill, but was an amusing and exciting pastime."[23]

not overemphasize the similarity between the game as it is now played and its early ancestors, but the major thing they have in common is the combination of chance and strategy. In the ancient games, the element of chance probably dominated, as it does in *El-tab el-siga,* which is still popular in Egypt today.

We will now turn to a bead game that better simulates natural chemical reactions than do the games we have just mentioned. In both Parcheesi and backgammon, the pieces can be moved in only one direction. Although the dice determine how far a player may move,

Table 7. BEAD GAME ''ELECTIVE AFFINITY''

Like Parcheesi, this game is designed for four players and uses a similar board. Each player has four beads. The game also requires two dice, a regular numerical cube and a die for making plus or minus decisions. For this purpose, a numerical die can be used. Even numbers mean "plus," and odd numbers, "minus." Or these decisions can be made by flipping a coin, heads or tails. The first die determines the number of spaces a player can move; the second determines the direction. Plus means clockwise; minus means counterclockwise. The players take turns rolling the dice and moving their beads. Whoever rolls a "six" is entitled to roll again. The decision on which direction the bead will be moved is not made until the number of spaces has been determined. Each player starts with one bead, which he places on the starting position closest to him. All the players then roll the dice once, and whoever gets the highest number of points begins the game. The players aim to reach any one of the four home bases with their beads. There is little point in a player choosing the home base closest to his starting point because he earns no points by doing so. It makes sense to choose this base only as a strategy to harm an opponent. A player earns the most points by placing his beads on the home base farthest away from his starting point. But because a player has to roll the dice often to reach that base, his beads will be exposed longer, and he will run considerable risk by not taking refuge at closer bases.

There is no obligation to choose any particular base. But whatever one a player chooses, he must roll the exact number needed to bring his bead to the farthest slot in the home base that is still empty.

This game is based on the cooperation of two players who join forces against the other two. A player's second, third, and fourth beads can be brought into play only with the help of a partner. This can happen only when two partners meet on one space, i.e., when one player rolls a space that his partner already occupies. The first meeting of this kind determines the partnership that will then hold for this entire game. The two remaining players must now try to bring about a meeting so that they, too, can play cooperatively. If one of them chooses instead to place a bead on a home base, he is then obliged to play out the game alone. In this case, he can count only his own points, and he can bring a new bead into the game only after he has placed one in play on a home base.

As a rule, of course, the players will try to form an alliance because cooperative play is advantageous to them. Whenever there is a meeting of partners, both of them can introduce a new bead into the game; and the player who brought about the meeting can also move six spaces in whichever direction he chooses.

If a player rolls a space occupied by an opponent, he has to pass. If two partners have all their beads in play, they can decide, when two of their beads meet, which of them will move six spaces. Because of this ruling, the game picks up momentum toward the end. It is over when all the beads of a single player or of two partners have reached home bases. Scoring is done according to the following scheme:

Eight points for any bead on the most remote home base.

Two points for any bead on the two intermediary bases.

Zero points for beads on the closest base.

Each team adds up its points, and each partner gets half of the total number of points. In the cooperative version of this game, it may sometimes pay, for tactical reasons, to occupy a base that brings in no points and so prevent an opponent from earning eight points.

---

these games still do not represent a true "random walk," which, even though limited to one dimension, has to be able to move backwards or forwards with equal probability. There are, of course, games of this type in which the pieces can be moved forwards or backwards, but these games are still not true examples of "random walk," because this choice is left up to the player.

Like Parcheesi, the game "Elective Affinity" is competitive. But the players can harm each other only indirectly. A somewhat frustrating game of "bad luck" is transformed into one of "good luck" which can be won by cooperation. The moves that are made before meetings take place and initiate reactions simulate a true one-dimensional "random walk" mechanism. If a player wants to reach the base farthest from him, he will have to roll the dice about four times more often than would be necessary if he settles for one of the two intermediary bases situated at half that distance. He must therefore carefully weigh the risk against the number of points to be won, but he should not exclude the possibility of good luck early in the game. He should also keep in

Table 8: BEAD GAME "STAY OUT OF 2D"

The name of this game comes from a motto with which Gerold Adam and Max Delbrück prefaced a study on the "Reduction of Dimensionality in Biological Diffusion Processes":[24]

DRUNKHARD: "Will I ever, ever get home again?"

POLYA: "You can't miss, just keep going and stay out of 3 D!"

As we show with our example of receptor effects in silkworms, the reduction of dimensionality, in that case from 3 D to 2 D, is of decisive advantage for efficiency of detection.

This game can be played by two, three, or four persons. On a playing board (see Figure 16), the following squares are distinguished from ordinary squares: the center square as starting position; four corner fields serving as home bases and consisting of four squares each.

Each player has four beads of one color. There are two tetrahedral dice whose surfaces are marked 0, 0, +, −.

The players begin the game by rolling the dice in turn. Whoever first rolls + + or − − opens the game by placing one of his beads in the starting position and rolling the dice once more. To move a bead one square as shown in Figure 16, a player has to roll either a plus or a minus. If a player rolls a + +, for example, he can choose whether he will move his bead two squares in the same plus direction or one square each in the two plus directions. The same principle applies to a roll of − −. If in the course of the game a bead returns to the starting position or if a bead is unable to leave that position at the beginning (because a player rolled 0, 0), that bead is returned to the reservoir. A player has to roll + + or − − again to place a new bead in the starting position, and whenever a player rolls + + or − −, he has to put a new bead in the starting position, provided that he still has beads in his reservoir and that the center square is free. He may then roll the dice again. This latter rule even applies if he has all his beads in play. Apart from these rules, it is left up to the player to decide which bead he moves. Once a bead has reached the squares in the border areas, it can be moved only on these squares. If a bead is moved onto a square already occupied by an opponent's bead, the opponent's bead is removed from the board, as in Parcheesi. If the dice select a square already occupied by a player's own bead, the player has to pass. The game is over as soon as any one player has placed all four of his beads on home bases.

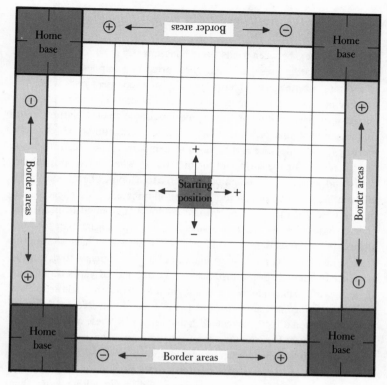

Figure 16. Playing Board for "Stay Out of 2 D."

mind that averages may not always be effective because statistical fluctuations can be considerable.

In nature, of course, the "random walk" motion is generally not restricted to one dimension but is two dimensional on a plane or three dimensional in space. In our next game we will see how the probability of meetings in the "random walk" process changes when we add new dimensions.

The crucial difference between the two games described here and popular board games like Parcheesi lies in their rules. In our bead games, the rules have been devised to simulate natural laws. That these rules produce new games is less significant than the fact that the results of these games reflect laws relevant to physics.

The diffusion game "Stay Out of 2 D" illustrates two important principles:

1. Most encounters between beads—and consequently most reactions—take place in the border areas, although these areas make up less than a third of the total playing area. The probability of encounters rises when the number of dimensions is reduced from two to one. The same principle applies if there is a reduction from three dimensions to two. Nature makes frequent use of this principle. The efficiency of a number of enzymes increases when they are arranged on a membrane. The substrate has a certain affinity for the surface of the membrane, where it can move about freely and where it can be guided directly to a point of reaction in the enzyme because one dimension has been eliminated. This phenomenon can be observed in particular with receptors that respond to individual effector molecules. As Dietrich Schneider[25] and his colleagues at the Max Planck Institute for Behavioral Physiology in Seewiesen have shown, the male silkworm in its moth stage makes use of this principle to detect the presence of a female over a distance of several hundred yards. Individual molecules of the pheromone the female emits to attract the male are picked up by the fanned-out antennae of the male and conducted by means of surface diffusion directly to receptors that react with these molecules. The stimuli so triggered are then interpreted to determine the gradient, i.e., the direction from which the pheromone comes.

2. The frequency of meetings is proportional to the population density on the board. That means that further meetings will occur with greater than average frequency immediately after a meeting has just taken place. This effect is not the result of any mysterious "law of series" but simply of a fluctuation that comes about because the density in one region of the board is temporarily increased by a meeting. In other regions of the board, of course, there is a corresponding decrease in density, and the frequency of meetings there is below average. It is easy to see from this how the frequency of meetings directly depends on population density. If reaction is triggered by meetings, the average rate of reaction will correspond directly to the average population density.

6.4 DISSIPATIVE PATTERNS

The course of every chemical reaction follows characteristic temporal and spatial patterns. But if a specific form is to take shape, specialized synchronization and self-regulation are necessary. Like selection, this process, too, requires a type of reaction capable of self-organization and including either feedback or autocatalysis.

As early as the turn of this century, Alfred J. Lotka and Vito Volterra studied the prototype of reactions that produce form. This prototype is based on a simple ecological problem which we will elucidate with a bead game. This is a "transformation game" that symbolizes both biochemical and ecological phenomena. This game closely simulates the changes that populations of different species undergo in the course of time and in a closed ecological environment. The scenario of the game is as follows:

1. Grass grows.	Green beads are placed on the board.
2. Rabbits eat the grass, and the rabbit population grows.	Green beads can be replaced by yellow ones, but only in areas where yellow ones are already present.
3. Foxes eat the rabbits, and the fox population grows.	Yellow beads can be replaced by red ones, but again only in areas where red ones are already present.
4. Foxes are hunted and their skins taken as trophies.	Red beads are removed from the board when they are "hit." These red beads represent the winnings of the individual players. They are exchanged for blue beads, which are in turn stored and counted up as points at the end of the game.

Table 9. BEAD GAME "STRUGGLE"

The requirements for this game are a square playing board divided into 64 squares; two octahedral dice whose surfaces are numbered with the coordinates of the squares on the board; 30 beads each of the colors green, yellow, red, and blue.

At the beginning of the game, 16 yellow and 4 red beads are taken from the reservoir of green, yellow, red, and blue beads. Keeping strategical considerations in mind, the two players place these beads anywhere they like on the board. "Strategic" means here that the beads should be so placed that the players can utilize the proximity of like-colored beads to bring about the most efficient possible transformations. In this game, proximity is defined as any of the four squares that are either vertically or horizontally adjacent to a central square. Once the 16 yellow and 4 red beads have been placed, actual play begins. The players take turns rolling the dice and replacing beads according to the rules summarized in the following chart:

square hit / adjacent square	empty	green	yellow	red
all squares empty	→ green	/	/	red → blue
green	→ green	/	green → yellow	red → blue
yellow	→ green	green → yellow	/	red → blue
red	→ green	/	yellow → red	red → blue

Note: At least one of the squares adjacent to a square chosen by the dice has to be occupied as shown in the chart. If there are beads of different colors in immediately adjacent squares, the player can choose which rule he will act on.

After a transformation as defined in the chart has been completed, it is usually the other player's turn to roll the dice. But this is not the case if further possibilities for reactions result from the initial transformation. Additional reactions can continue to take place as long as they conform to the chart. Only after these possibilities are exhausted can the other player roll.

A chain reaction can take place if a transformation produces the neighboring color combinations green-yellow or yellow-red. These combinations are then transformed as follows: green-yellow becomes yellow-yellow; yellow-red becomes red-red.

This transformation rule for chain reactions applies whenever the combinations green-yellow or yellow-red occur, regardless of which of the two beads was exchanged in the preceding transformation. A new proximity is always defined in terms of the bead that has just been transformed.

For every transformation of yellow-red to red-red, a player gets two extra rolls of the dice. But the player can act on these rolls only if the dice select a red bead. If a red bead is chosen, it is removed from the board, exchanged for a blue one, and counts, in this form, as a point.

The game is over when either all the yellow or all the red beads have disappeared from the board. Whoever has the most blue beads is the winner.

The rules of this game simulate in concrete form the four stages of the reaction listed in the scenario preceding this table.

The filling of empty squares with green beads represents the growing of grass. We are not introducing an autocatalytic process here because we can assume that grass seeds are present in the soil in sufficient quantity.

But the next two reactions, being directly dependent on a finite population of the species in question, are autocatalytic in nature. If no rabbits or foxes are present, no rabbits or foxes can be born. Furthermore, an increase in both species depends on the availability of food: Rabbits eat only grass; foxes eat rabbits (but no grass). The shooting of foxes is not an autocatalytic process, because the number of hunters in a given area will not depend on how many foxes have been killed there. The more foxes there are in the area, the more that will be shot, but this does not necessarily mean that the number of hunters has increased.

This kind of ecological reaction scenario illustrates in concrete terms the general mechanism that Lotka and Volterra investigated and analyzed in mathematical terms. They summarized their findings in this abstract formulation:

$\rightarrow A$	Green beads (A) are placed on empty squares.
$A + X \rightarrow 2X$	Green beads (A), with the help of yellow beads (X), are transformed into yellow beads (X).
$X + Y \rightarrow 2Y$	Yellow beads (X), with the help of red beads (Y), are transformed into red beads (Y).
$Y \rightarrow B \rightarrow$	Red beads (Y), in proportion to the frequency of their occurrence, are transformed into blue beads (B) and are removed from the board.

The reaction begins when the basic substance A comes in contact with its reaction partner X and is transformed into X. If the autocatalytically functioning partner X (the rabbit) happens upon a local concentration of the substrate A (grass), this concentration will be reduced, but the growth rate of X rises as a result of autocatalytic reproduction. The growth rate will drop after some delay, i.e., as the supply of A becomes scarce. In the meantime, the local population density of X has risen well above the average and will decline again only when the supply of A is depleted. If a second autocatalytically functioning reactant Y (fox) that transforms X into Y is present, then the supply of X will rapidly decrease as Y increases, and Y will go through the same fluctuation of increase and decline that X has. X will soon sink way below an average local population density, and A will have a

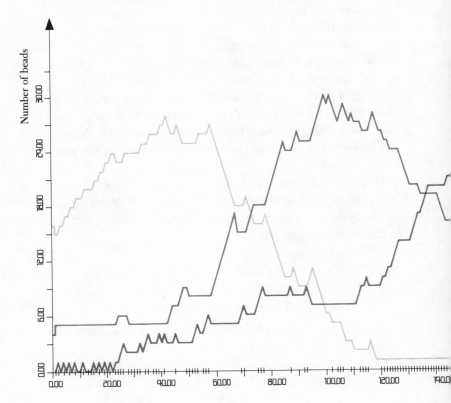

chance to grow again undisturbed. The whole cycle then repeats itself. This process resembles a periodic overflow from one reservoir into another. Figure 17, which graphically represents this game as played by a computer, illustrates this effect. We will examine another highly instructive example of a periodic reaction in Chapter 12 (see "Hypercycle," Table 14). Given the proper conditions, a temporally periodic process can easily produce a spatially periodic pattern.

In an initially homogeneous medium A a local disturbance— "inoculations" with X and Y—will spread in a wavelike pattern similar to the one that occurs when we throw a stone into calm water. The stone thrown into the water demonstrates how—for a certain period —spatially stable patterns such as standing waves can form. Spatially stable patterns arise in a similar way in an oscillating reaction system

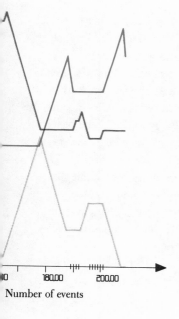

Number of events

Figure 17. "Struggle." Graph of this game as played by a computer. The temporal changes in the different populations are reflected in the alternating peaks of their oscillations. The individual components are represented by the colors of the curves (green = grass, yellow = rabbits, red = foxes).

if there is a constant supply of the reaction partner (A, for instance) and a diffusion and removal of the products of the reaction (primarily Y). These patterns are called "dissipative structures" because energy has to be dissipated steadily to maintain them. If we ask now what gives rise to spatial and temporal order, the answer would be that it results solely from the properties of the reactions inherent in the system and from the boundary conditions that apply to the system. This order is maintained by a steady flow and dissipation of energy, i.e., by the intake of substances rich in energy and the discarding of substances low in energy.

This process represents a second fundamental principle of order in nature. Dissipative structures are just as significant for the production of form in animate nature as is conservative order, which is based solely on static forces. It is possible to formulate precisely in mathematical terms what minimal conditions a reaction system has to fulfill to produce a dissipative pattern. There are no limits to the complexity of structures occurring in nature. The spatial patterns present in the conservative structures of proteins are no more intricate than those in the wave and oscillation pattern that the individual components of a multiphase reaction cycle go through.

Many physicists, chemists, and biologists are at work today trying to understand the secret of biological form. In the 1950s, the English mathematician Alan M. Turing discovered the significance of autocatalytic reaction mechanisms for morphogenesis. The Belgian physical chemist Ilya Prigogine[26] made some basic theoretical studies in this area. René Thom's[27] research, which inspired a renaissance in the mathematical discipline of differential topology and laid the basis for what is now known as "catastrophe theory," provided new insights into the problem of biological form. Figures 18 through 21 illustrate some examples of dissipative structures in inorganic chemistry and biochemistry as well as in cell biology and neurobiology.

Let us now summarize what we have learned about biological form up to this point. There are two fundamental principles of morphogenesis, a conservative one and a dissipative one.

According to the first principle, structure and form result from an interaction of the conservative forces of attraction and repulsion. The

Figure 18. Dissipative patterns caused by an inorganic chemical reaction (Zhaboutinsky reaction). This process involves autocatalytic stages without which oscillation could not take place. The illustrations, made in the course of an experiment conducted by Benno Hess at the Max Planck Institute for Nutritional Physiology in Dortmund, show clearly how a reaction, starting from two centers, spreads in waves and, because of interference, forms a complex pattern. This pattern remains stable only for the duration of the reaction, i.e., only as long as energy is being dissipated. It disappears as soon as the substances rich in energy are used up and the reaction comes to a halt.

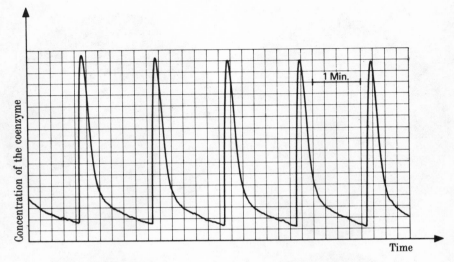

Figure 19. Periodic time pattern of a biochemical reaction as recorded by Benno Hess.[28] This is the pattern of one phase in the enzymatic reaction that takes place when sugar is broken down by glycolysis. The concentration change of the coenzyme was recorded spectroscopically. The reaction was kept going by a constant input of glucose.

subunits of the overall system, subject to the permanent action of these forces, assume stable spatial positions or move in stable orbits (as the planets do) around a focal point. This order is maintained *without* a dissipation of energy.

Dissipative structures differ from conservative ones in being dynamic states of order that can be maintained only by means of a metabolism, i.e., a constant dissipation of energy. From the combination of the transport of matter and of synchronized, periodic transformations, they form spatial patterns similar to standing waves. They cannot be produced simply by adding together their subordinate structures. In morphogenesis, dissipative structures are responsible for determining and spatially organizing the elements of conservative structure, specified by the genetic program of the cell. As patterns of stimuli in the network of the nerve cells, dissipative structures represent more than the sum of the informational elements they contain and are therefore a material correlative of gestalt. The interactions necessary for the formation of dissipative structures are based on conservative forces, just as the spatial constancy of dissipative patterns requires stabilizing, conservative forces.

What the conservative and the dissipative generation of form have in common is cooperativity in static or dynamic interactions. The abstract presentation of these principles in our bead games has demonstrated this point: *The cooperative interplay of forces in conservative patterns corresponds to autocatalytic reactivity in dissipative models.* In the case of conservative cooperativity, an empty square selected by the dice will be filled with a bead, and this bead will have the same

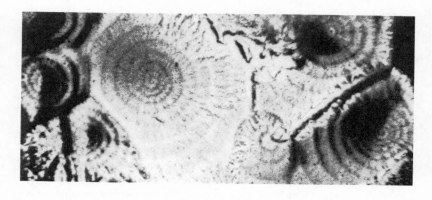

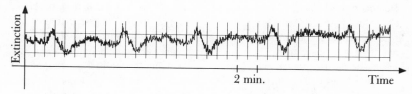

Figure 20. Invisible forces or a "god of amoebae" seems to cause single-celled slime molds to combine in a multicellular plasmodium that then acts like a single organism. Günther Gerisch[29] of the Friedrich Miescher Laboratory in the Max Planck Institute in Tübingen studied this phenomenon and discovered that it is caused by a dissipative process based on an autocatalytic reaction mechanism. The photograph shows the dissipative pattern of stimulation in a cell population; and the graph below, made by Benno Hess, shows the periodic time pattern with which the chemical signal initiating an aggregation is transmitted.

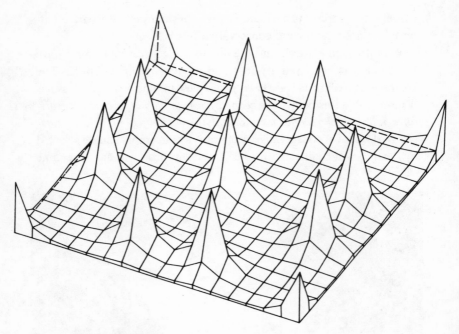

Figure 21. Computer drawing of a dissipative spatial pattern as calculated by Alfred Gierer and Hans Meinhardt on the basis of the model shown in Figure 14. Activated regions form in a concentration distribution assumed to be homogeneous. Jack Cowan and Hugh R. Wilson[30a] as well as Christoph von der Malsburg[30b] have found similar patterns operative in the activation of nerve-cell networks (see p. 297). The psychological concept of "gestalt" is reflected here in material form. A dissipative gestalt results from the superimposition of a number of individual processes, including the feedback process, on each other so that the final result is more than the sum of the parts.

color as a bead already present in an adjacent square. In an autocatalytic reaction, on the other hand, an empty square adjacent to an occupied square selected by the dice will be filled with a bead of the same color as the one in the occupied square. In both cases, two adjacent squares, only one of which was originally occupied, end up with two beads of the same color.

We can best illustrate the differences between these formative mechanisms by comparing the mechanistic details involved.

1. In the dissipative model, a constant pattern takes shape without the particles being spatially fixed in a reproducible way.
2. In contrast to the conservative model, dissipative form is not determined solely by the interaction taking place between material particles but is decisively influenced by the boundary conditions and limitations of the system.
3. The maintenance of dissipative structures requires—as the name suggests —a continual dissipation of energy, which means there is a constant production of entropy (cf. p. 141). In other words, the system has a metabolism. Free energy supplied by the transformation of matter is continually used up.
4. Conservative structures have a higher degree of "absolute" stability, reversibility, and superposability. That is, these factors are less affected by external constraints. But because dissipative structures are dependent on such constraints, they can be combined or superimposed on each other only within certain limits.

Goethe, with his keen powers of observation, did not fail to pick up this last point, although his classification according to different types of form was done purely in phenomenological terms and was therefore somewhat arbitrary. As he remarked, if the "forms [of organic beings] are destroyed, they cannot be reconstructed from what remains," whereas a "mineral we thought destroyed can soon be restored to its original perfection." In this inherent tendency of minerals to form structures, Goethe detected an "indifference their components show toward the form of their combination," and he ranked the organization of inorganic structures below the perfect totality found in organic ones.

From the perspective of current research, the problem of organic form can be understood only in terms of the interplay between the conservative and the dissipative principles. Dissipative processes direct and synchronize how information stored in conservative structures will be elicited from them and guarantee the functional effectiveness of that information. The fact that spatial and temporal patterns can be translated into the abstract language of an informational program is evident in the material self-organization of living organisms as well as in the make-up of our ideas. Indeed, what the concept of "form" or "gestalt" tries to describe originates in our perception. Wolfgang Köhler in particular used methods of experimental psychology to demonstrate this principle underlying our patterns of perception and thought.[31] Through his research, he discovered a principle whose broad applicability as a model we have only begun to work out today, as in our model of how dissipative structures aid in the integration of conservative interactions.

7

Symmetry

In his book Symmetry, *Hermann Weyl wrote: "As far as I can see, all a priori statements in physics are based on symmetry." The Platonic concept postulates symmetry as the ultimate basis of relationships. Breaches of symmetry represent gaps in our understanding of fundamental connections. In the structures of reality, symmetry is revealed a posteriori. In animate nature, as well as in games, symmetry occurs only when it is associated with selective advantages.*

7.1   THE PLATONIC CONCEPT

A treatise on form that omitted symmetry would be like a tour of the art centers of Italy that bypassed Florence. There are few phenomena that have fascinated thinkers so much as symmetry. "At the heart of all phenomena," Werner Heisenberg wrote, "we find the mathematical law that defines basic symmetrical operations, such as transpositions in time or space, and thereby determines the framework in which *all events* occur."[32] But the forms and shapes of the world we perceive

Figure 22. A small selection from the 2,453 photographs of natural snow crystals assembled by W. A. Bentley and W. J. Humphreys in *Snow Crystals* (Dover, 1931).

with our senses are by no means predominantly symmetrical. Indeed, even in crystals, which are considered to be the epitome of symmetrical form,* symmetry is subject to the vagaries of chance. This point is illustrated in the incredible variety of forms that snowflakes take (Figure 22):

> Little jewels, insignia, orders, agraffes—no jeweller, however skilled, could do finer, more minute work. . . . And among these myriads of enchanting little stars, in their hidden splendor that was too small for man's naked eye to see, there was not one like unto another.[33]

Thomas Mann's profound appreciation of this "infinite inventiveness in the variations and delicate execution of one and the same pattern" struck a responsive cord in Hermann Weyl, one of the great figures of the Princeton era. In his book *Symmetry,* [34] Weyl offers deep insight into this concept, its foundations in group theory, its applications in physics, chemistry, and biology, and its role in art.

The simplest form of symmetry is based on mirror images on either side of a straight line. This is the "right-left" or bilateral symmetry that occurs with particular frequency among higher organisms. It has been used in art in countless variations from the Neolithic Age of Catal Huyuk (Figure 23) on down through Babylon, Egypt, Greece, and Rome into the modern era, where we find it, for example, in the periodic drawings of Maurits Cornelius Escher (Figure 24).

The second important means of creating symmetrical forms is the progressive spatial repetition of a basic pattern. This process can consist of a translation or rotation through an angle equal to 360 degrees divided by some integer. All symmetries can be derived from

---

*Statistical physics often uses the concept of symmetry in a way that could easily lead to misunderstandings in the present context. If a statistical distribution is characterized by the absence of spatial correlation, it is called perfectly symmetrical. Consequently, only a "breakage of symmetry" in the homogeneous distribution could produce a crystal with its distinct intervals and orientations. The perfect symmetry of a random distribution can, of course, be fulfilled only in terms of *temporal* or *spatial averages.* Any individual distribution at any given time would be highly "asymmetrical," whereas the individual distributions in a crystal always conform to the iteratively symmetrical order of its building blocks.

a combination of these two basic operations. The playing board provides us with a ready means of studying the formative laws affecting terative two-dimensional patterns (see Figure 25 for an example).

The original meaning of the Greek word *symmetros* is "regular," "well-proportioned," "harmonious." As the last of these synonyms suggests, the word "symmetry" did not refer exclusively to phenomena limited to geometric space, and today we use the term primarily in an abstract mathematical sense. But this use of the word, too, goes back to the Greeks' observations of nature.

The early Greek settlers in southern Italy were familiar with the form of the dodecahedron, probably from their observations of pyrite crystals. The surfaces of the pyrite crystal itself, however, do not take

Figure 23. Bilateral symmetry. Plaster relief of a pair of leopards on the north wall of a temple in Catal Huyuk, ca. 5800 B.C. (James Mellart, *Catal Huyuk: A Neolithic Town in Anatolia* [McGraw-Hill, 1967]).

Figure 24. "Smaller and Smaller" (1956) by Maurits Cornelius Escher (1898–1972) (*Die Welten des M. C. Escher* [*The Worlds of M. C. Escher*], Heinz Moos Verlag, Munich). This illustration is used with permission of the "Escher Foundation," Haags Gemeentemuseum, Den Haag.

the form of regular pentagons, because this symmetrical shape does not conform with the laws of crystallography. Hermann Weyl called the discovery of the regular solids one of the most wonderful unique abstractions in the history of mathematics. How unique this discovery was is debatable. The Etruscans, too, were familiar with the dodecahedron, which they used as a religious symbol.

We associate the theory of regular solids primarily with the Platonic school. Theaetetus, a student and friend of Platos, was responsible for working out this theory in detail. Plato himself saw the regular solids embodying the basic elements of a geometrical theory of matter, a theory he sketches briefly in his dialogue *Timaeus*.

Plato associated the tetrahedron, cube, octahedron, and icosahe-

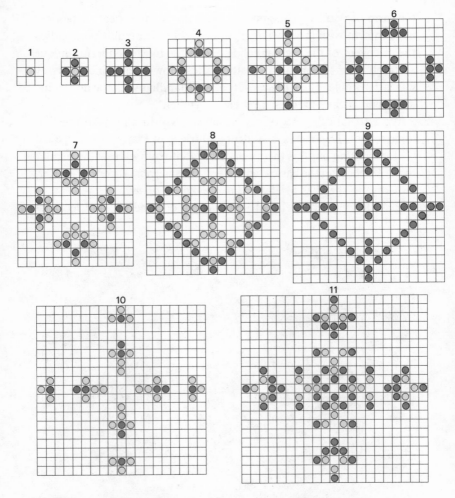

Figure 25. Symmetrical patterns arise in this reproduction game worked out by Stanislav Ulam. In each generation, those squares in whose orthogonal neighborhood (i.e., the four orthogonally adjacent squares) one and only one bead is already present are filled. All the beads of a generation (n) die out as soon as the generation (n + 2)

appears. This means that at any given time only the beads of the last two generations are on the board. The illustration shows the development of a pattern that arose from a single bead. The final pattern represents the forty-fifth generation.

dron (Figure 26) with the four elements of fire, earth, air, and water. The dodecahedron symbolized the world as a whole, probably because Plato could not find an appropriate simple correlation for it. (At least his explanation "that God used the pentagonal surfaces of the dodecahedron to place figures on" is not very convincing in terms of geometrical theory.) A further breaking down of the four basic elements destroys their nature as "elementary bodies." According to Plato, such a breaking down creates the abstract, two-dimensional forms of the polyhedral surfaces. If the tetrahedron, octahedron, and icosahedron are taken apart, congruent equilateral triangles remain. Plato regarded this congruence as an explanation for the possible mutual transformation of the "elements" of fire, water, and air. In the case of the cube, of course, a further symmetrical division of the six

Figure 26. The Platonic Solids.

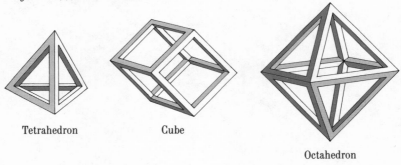

Tetrahedron   Cube

Octahedron

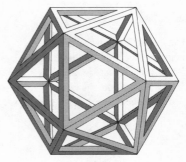

Dodecahedron   Icosahedron

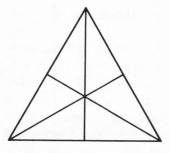 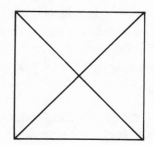

Figure 27. Plato's Basic Triangles

square faces does not produce equilateral triangles. Plato therefore abandons the idea of further division and defines as the "most beautiful"—and therefore irreducible—basic forms those triangles that include a right angle, i.e., triangles resulting from the symmetrical division of an equilateral triangle as well as of a square (Figure 27). Here, too, the cube occupies a special position. Plato interprets this to mean that earth in its pure form cannot be reduced to the other elements, which can be mutually transformed. "Water" can be turned into "air" if we interpret air as a collective term for vapors or gases, and air can become fire if we define the hot, flaming gases that escape from burning material as fire. Plato also assigned the hierarchical structure of the polyhedrons a special significance in view of the differing degrees of volatility the four elements had.

Plato admitted the arbitrariness of his classification by triangles, under which the dodecahedron could not be subsumed, and he said:

> If anyone can name a more beautiful triangle underlying the composition of solids, we will greet him not as an opponent but as a friend in the right.[35]

Why have we given so much attention to a view of nature irrelevant in terms of modern science? What we want to show—and the remark we have just quoted emphasizes this—is that Plato was not intent on reducing matter to specific elementary units that could be described in detail, but was instead searching for a method based on mathematics (and harmony) that could explain all interrelations and thereby give a unified description of the world. This is the approach that Plato and modern physicists share.

The search for the fundamental units of matter, a search that broke matter down into even smaller particles, left physicists facing a dilemma. The more closely they studied matter, the more so-called elementary particles they found. They had hoped, by constructing increasingly large and complex apparatuses, to get a glimpse of the mysterious "elementary substance." But what they found instead was a large number of new, short-lived particles. Consequently, one of the crucial tasks of modern physics is to introduce order into the wealth of newly won experimental insights, and scientists are at work today trying to discover the basic symmetries that account for the irreducible forms in which matter manifests itself.

Since ancient science did not make use of experimentation, it had to rely solely on the formulation of a priori postulates, and the remarkable thing about Plato is that he did not look for the key to nature's mysteries in physical details themselves, as Democritus before him and many others after him did, but sought it instead in the relationships, reducible to màthematical formulation, that underlay those details. It is idle to wonder whether Plato was conscious of how radically his approach differed from that of other scientific thinkers. What is clear, as we saw in the quotation above, is that he regarded his theory of the elements as no more than a preliminary sketch, and he assigned it only a subordinate place in the body of his work. As Plato's concern with finding the "most beautiful" triangles indicates, he gave great weight to the aesthetic point of view. More than two thousand years would have to pass before this kind of inductive description of nature could be stripped of its arbitrary quality. Before that could happen, scientists had to realize that the final proof for any a priori assumption can be found only in the experimental testing of its consequences, indeed, that any knowledge we gain will be valid only in the realm in which it is gained. To Wittgenstein's apodictic statement, "Everything that can be said at all can be said clearly," we must therefore add, "only within limits that can be experimentally verified." All the basic theories of modern science are essentially offshoots of new and unexpected results obtained in experiments.

Die Gleichen.

Figure 28. Die Gleichen ("The Equals")
near Göttingen.

7.2 BROKEN SYMMETRIES

Near Göttingen are two hills that are called Die Gleichen—"The
Equals" (Figure 28). They were no doubt given this name because,
when viewed from a certain angle, they look almost identical. It is
reported that David Hilbert used to ask his students why these hills
were so called. None of the obvious answers like "equal height" or
"equal shape" satisfied this great scholar, who was a stickler for
exactness. Finally he would give the answer himself: "Because they
are equal distances apart from one another."

Mathematicians were the first to doubt the general validity of
knowledge gained through observation. Carl Friedrich Gauss did not
take for granted that the laws of Euclidian geometry held for all
magnitudes, and he set about testing them experimentally. But the

triangle he laid out between three mountains—the Hohe Hagen near Göttingen, the Inselsberg in the Thuringian Forest, and the Brocken in the Harz Mountains—was by no means large enough for the sum of its angles to show even the slightest deviation from 180 degrees. Gauss's doubts, nonetheless, soon proved to be justified. His successor, G. F. Bernhard Riemann, pursued Gauss's ideas further, and it was Riemann's work in spatial geometry that laid the foundations for the general theory of relativity. Ever since then, physicists, too, have made it a principle to be suspicious of "obvious" facts, and there are no cows that remain sacred for them anymore.

Only a few years ago it was discovered that an assumption about symmetry that had been considered self-evident was not in fact true, in one special case at least. This discovery caused great excitement, for none of the previous "revolutions" in this field had called the old principles of symmetry and invariance into question; to the contrary, they had tended to emphasize the importance of these principles. The relativity theory developed by Albert Einstein, Hendrik Antoon Lorentz, and Hermann Minkowski, for example, is essentially a theory that tries to preserve symmetrical relationships when phenomena are represented in a four-dimensional space-time continuum.

Subsequent research into the world of elementary particles at first uncovered nothing but additional symmetries. The most important of these was the discovery that for every elementary particle there is a "material mirror image." In this mirror image, all the signs are reversed. The positron, the counterpart of the electron, is, as its name suggests, positively charged, but its mass corresponds exactly to that of the electron. Neutral particles have their antipodes, too. For them as well, all properties that have signs, like magnetic moment, for example, are reversed. The English physicist Paul Dirac predicted the existence of such "holes in matter" or "antiparticles" as early as 1928. In 1932, they were discovered in cosmic radiation, but they could not be reproduced in the laboratory until the large accelerators of recent years were developed. Antiparticles are created by an energy-rich collision of highly accelerated matter, as are their mirror images, provided that certain quantum conditions are fulfilled. Conversely, they both destroy each other's material existence whenever they meet, and in the process they give off the same amount of energy as was needed to create them. There is, then, a good reason why the mirror image of material particles cannot ordinarily be observed. A world of antimatter is conceivable, but it could not coexist with our world made up of "normal" matter.

An electric charge consequently is a property that can change its sign by means of a mirroring, which we call a "C" ("charge") transformation. Are there other similar symmetrical operations? From mechanics we know—and modern amplifications of this principle have not affected its basic application—that a mirroring on a time axis, i.e., a reversal of past and future, is accompanied by a reversed sign for the direction of motion but that the *laws of motion* remain the same.

Another very important symmetry, invariance under reversal of all three spatial coordinates, led to far-reaching new insights. Invariance under spatial inversion implies conservation of a quantum number called parity. Before 1956, very few physicists doubted the absolute validity of parity conservation—it was one of the sacred cows. In that year, however, two Chinese physicists living in the United States, Tsung Dao Lee and Chen Ning Yang, on the basis of inexplicable experimental findings, such as the so-called tau-theta puzzle, wrote a paper in which they questioned parity conservation in a particular type of forces (the weak interactions) and suggested ways to test their revolutionary hypothesis. We will briefly describe one of the proposed experiments here.

A breakdown of parity implies the existence of a left-right asymmetry in the decay of atomic nuclei. In order to search for such an asymmetry, atomic nuclei must be polarized so that their spins all point along the same axis. Every nucleus is like a small, continually rotating top, which has, because of the rotation of electric charges, a magnetic moment and therefore represents a magnetic dipole. In a strong magnetic field at very low temperatures, all dipoles point in the same direction; the nuclei are polarized and rotate in the same direction; their spins are parallel. If the left-right symmetry is not violated, and if the polarized nuclei emit electrons, as many electrons must be emitted in the direction of their polarization ("up") as opposite to it ("down"). Conversely, if the ratio up/down differs from one, parity is violated.

The experiment proposed by Lee and Yang was carried out by Madame Chien Shiung Wu and her colleagues using cobalt 60. This isotope is radioactive, and when it decays, it emits an electron, a negatively charged beta particle. This process is characteristic for "weak interactions." The experiment was designed to test parity conservation in transformations induced by weak interactions. The result of this experiment, which had truly sensational consequences, was that electrons left the polarized nuclei predominantly in one direction (see Figure 29), thus demonstrating an "up-down" asymmetry. The interaction underlying beta decay therefore had to have an inherent direction of spin. We must remember, of course, that we are referring here only to the so-called weak nuclear forces studied in this experiment. Further experiments showed that the direction of spin was reversed if a particle was replaced by its mirror image, an antiparticle, whose charge is reversed.

An experiment always represents a question we are putting to nature. In the case of the parity problem, the answer was unequivocal. Physicists had to accept it as a given. (The story is told that a supremely self-confident theoretician, responding to the objection "What if the experimental facts contradict the theory?" replied, "Too bad for the facts!")

These experiments became more fascinating still as scientists began to combine different symmetry operations. Apart from the three mirror images for C, or charge, for P, or parity, and for T, or time, they now examined combinations CP and CPT more closely.

It had been assumed that natural laws would hold unchanged for mirror images. The Lee, Yang, and Wu experiment showed that this was not the case for the mirror images of P and C, that is, for reversals of parity and charge, in weak interactions. But Gerhart Lüders, Wolfgang Pauli, and Julian Schwinger had previously developed a theorem stating that for the *combination* of C, P, and T, natural laws retained their validity under all circumstances. This so-called CPT theorem had been confirmed by experience.

Another implicit assumption that followed on the discovery of breaches of symmetry in C and P was that these breaches occurred in pairs (in both C and P, for example) but were then compensated for in the combination CP. The mirror imaging of T therefore had to be an operation that left natural laws intact.

But an experiment conducted in 1964 came up with the astonishing result that the combination CP also contains a breach of symmetry. If the CPT theorem is to remain valid—and there is no reason to call it into question—then the mirror imaging of time in weak nuclear forces must also include a breach of symmetry. What was new and at odds with previous knowledge was the fact that invariance did not result from the *combination* of the two symmetry-violating operations C and P. These findings clearly demonstrate that the theory is still inadequate and that there is probably a new and yet to be discovered interaction that underlies this incomprehensible fact.

This digression may have taken us further afield in the esoteric reaches of physics than we meant to go. But it has enabled us to show that no theory stands beyond experience. Indeed, observation of facts always precedes understanding. This is as true for the expert as it is for the general reader.*

The problem we have just described is symptomatic for the situation of modern physics, and it represents only a glimpse of the manifold mysteries in this field that are still waiting to be solved.

At the moment, the theoreticians are divided into two camps. The optimists continue to subscribe to the Platonic ideal and to search out a new, higher level of reality on which broken symmetries can be pieced together again. Making use of a special interrelationship between symmetries that Wolfgang Pauli has discovered and of a new metric, developed by Paul Dirac, Werner Heisenberg has been the most ready to speculate on the universality of theory. Heisenberg claims to see a portrait of "unity in nature" emerging; the great skeptic Pauli, on the other hand, accepted no more than the frame around this portrait.

Scientists in the other camp—and we do not necessarily want to

---

*Detailed descriptions of the experiments mentioned here and of their theoretical implications can be found in Frauenfelder and Henley's *Subatomic Physics.* [36]

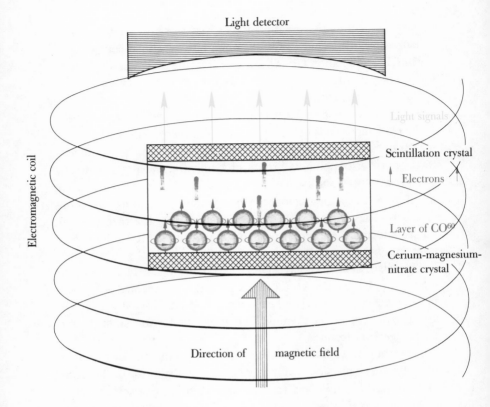

Light detector

Light signals

Scintillation crystal

Electrons

Layer of CO⁶⁰

Cerium-magnesium-nitrate crystal

Electromagnetic coil

Direction of   magnetic field

Figure 29. The breakdown of parity in "weak interactions" was proved in a "classic" experiment conducted by Madame C. S. Wu of Columbia University and E. Ambler, R. W. Hayward, D. D. Hoppes, and R. P. Hudson of the National Bureau of Standards.

A thin layer of the isotope cobalt 60 was crystallized on a layer of cerium-magnesium-nitrate crystal. A scintillation crystal was mounted about two centimeters away from the cobalt layer. When electrons struck this crystal, it emitted flashes of light. The light flashes emitted were received and amplified by a photodetector outside the electromagnetic coil and the cooling system. The sample to be tested and the scintillation crystal are located inside a coil in which an electric current creates a magnetic field. The sign of the field can be reversed by

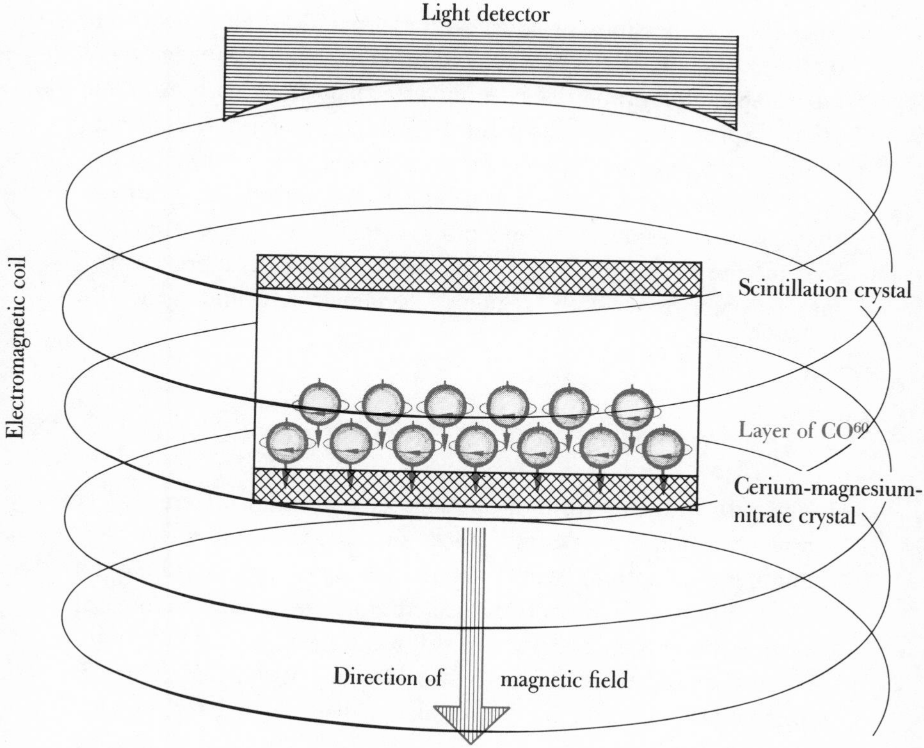

Light detector

Electromagnetic coil

Scintillation crystal

Layer of CO^{60}

Cerium-magnesium-nitrate crystal

Direction of magnetic field

changing the direction of current flow. The entire apparatus for the experiment is contained in a cryostat. The cerium-magnesium-nitrate crystal was cooled to near absolute zero by adiabatic demagnetization. The cobalt nuclei were polarized in the magnetic field, and all the atomic axes therefore point in nearly the same direction, i.e., the nuclei all rotate in the same direction.

The experiment shows that electrons released during beta decay do not leave the polarized cobalt nuclei in equal numbers toward both poles. The differential signals, which depend on the polarization of the nuclei, occur only under the experimental conditions shown in the diagram on the left.

label them as pessimists—agree with Parmenides that "everything that is exists only in the idea that comprehends it." For them, there are no absolute symmetries in a universe that, along with its whole metric expressed in the basic natural constants, is subject to continual change. These two convictions reflect complementary aspects of one and the same reality. The task at hand is to define the level of thought that, given the structure of our brains, yields the most comprehensive picture of reality; and in the last analysis, "comprehensive" may be synonymous with "regular," "symmetrical," and "generally applicable."

7.3 SYMMETRY A POSTERIORI

Let us return to the world of our sensory perceptions. Here perfect symmetry is almost unknown. Most shapes and forms in our surroundings have nothing regular about them. In modern art, almost all directly perceptible proportion has disappeared. Architecture, of course, represents an exception in still using the cube as a basic unit. Symmetry is somehow characteristic of living organisms, though it appears more often in higher forms than in lower and in later ontogenetic stages rather than in very early ones. The form of an individual cell is fluid; the proportions of a Venus can be captured in marble.

Symmetry in nature is the result of an evolutionary process, never the cause. Symmetry has to embody a selective advantage; otherwise it could not hold its own in the interplay of chance and necessity, of mutation and selection. Life plays with symmetry the way a composer plays with rhythm and harmony. Two examples from molecular biology illustrate this relationship between selection and symmetry. The first example shows that nature will not make use of a symmetrical solution unless it proves more advantageous than an unsymmetrical one. Here too, nature seems to violate parity, but the analogy to the violation of parity in subatomic physics we just described is only formal.

In the mid-nineteenth century, Louis Pasteur discovered that the metabolism of living cells clearly prefers to make use of compounds that are characterized stereochemically by one direction of rotation. We could express this by saying that the metabolic enzymes work with only one hand. And chemistry does in fact use the word "handedness" or "chirality" (from the Greek *cheir,* meaning "hand") to describe this stereochemical phenomenon. What causes this chirality?

Also in the nineteenth century, Jacobus Hendricus van't Hoff found that the carbon atom has a tetrahedral structure and that it forms an asymmetrical center when it is joined by four different ligands. This asymmetry can be detected with the help of polarized light. The reader unfamiliar with this concept need only know that there is an experimental device that shows:

- whether the substance under examination is optically active, i.e., whether it contains an asymmetry, for only in that case will the needle of the instrument be deflected at all
- what the chemical compound's direction of rotation is, i.e., whether it turns light to the right or the left. The needle reflects the direction of rotation by moving either right or left
- how strong the optical asymmetry is. This is indicated by how far the needle moves on the numbered scale of the instrument.

But because the tetrahedral corners of a carbon atom are equal in value, an asymmetry is present only when all four ligands are different from each other. Figure 30 illustrates this. Only if the four ligands are different will the mirror image no longer correspond to the image produced by rotation of the molecule. The two positions, before and after rotation, will then correspond to values of optical rotation that are identical in absolute value but that have different signs.

The building blocks of one of the most important classes of biological macromolecules, the proteins, contain asymmetrical carbon atoms of this kind. All the protein building blocks in the entire organic world, ranging from coli bacteria on up to human beings, are left-handed. This means that any macromolecular chains built up from them will reflect this uniform direction of rotation. The proteins form spirals that turn to the left; these are the alpha helices predicted by Linus Pauling. This does not mean, however, that the mirror images of the building blocks are unknown to chemists. In test tubes, they are

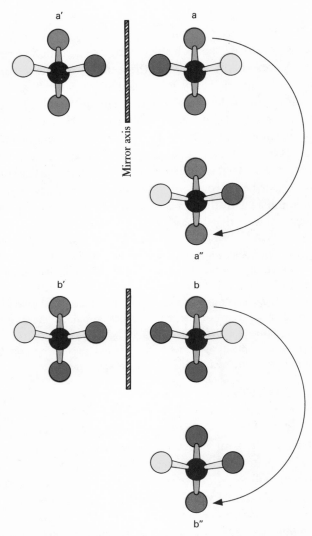

Figure 30. A chiral center in a carbon atom can exist only if all four ligands are different. If two ligands are the same (upper half of illustration), a rotation of 180 degrees around an axis perpendicular to the page and passing through the black carbon atom results in an arrangement of ligands (a″) that corresponds exactly to that of the mirror image (a′); (a″) is indistinguishable from the mirror image (a′). If all four ligands are different (lower half of illustration), the mirror image (b′) and the ligand arrangement resulting from rotation (b″) are no longer identical.

produced just as often as the left-handed blocks are. Indeed, at lower molecular levels, they can even be produced and utilized by living organisms. But they are never incorporated into the molecular chains of proteins. They have to be built up by special enzyme machinery (cf. Chapter 15) that is in no way related to the biosynthetic mechanism that produces proteins. With the proteins, only left-handed rotation occurs. But with the nucleic acids, just the opposite is true. Here, right-handed rotation was nature's choice.

What accounts for this broken symmetry?

It is easy to see that the use of both directions of rotation in the same macromolecule would be disadvantageous. Evolutionary processes resemble political elections in that one side has to win; and something like "team unity" in the process of synthesis is advantageous. In chains with mixed chirality, the machinery of synthesis would have to be very complex to adjust itself constantly to incorporating left- and right-handed elements. An efficient enzyme is exactly suited stereo-chemically to its substrate. We could hardly expect a primitive molec-ular system to be so versatile as to be able to keep making new stereochemical adjustments to different substrates.

But why then do we not find in nature two kinds of protein or nucleic acid molecules that coexist peacefully, one kind made up exclusively of left-handed building blocks and the other made up exclusively of right-handed ones?

Once upon a time there was a scorpion who wanted to cross a river. A turtle happened to come along, and the scorpion asked the turtle to take him across the river.

The turtle answered, "But how do I know that you won't sting me when we're halfway across?"

"Don't be silly," the scorpion replied. "If I did that, we'd both drown."

The turtle let himself be convinced, but when he had reached the middle of the river, the scorpion stung him after all. As he was dying, the turtle asked, "How could you do this to me?"

And just before the scorpion himself went under, he replied, "That's my nature."

Similarly, it was the nature of early self-reproducing molecular systems that prevented the peaceful coexistence of right and left. Many biologists regarded the breach of symmetry in the chirality of macromolecules as a proof that the origins of life on our planet could be ascribed to one unique event. But the evolution of life, especially in its early stages, is the result of reproduction, mutation, and selection. It is therefore irrelevant how often life-originating events took place. Because of the "once and for all" quality of selection in this phase (cf. Chapter 12, p. 232), only *one* of the possible alternatives could prevail. Which of the two variants was selected was a matter of chance. In our bead game "Selection," too, we cannot predict in advance who the winner will be. All we can say is that even if two players start with exactly equal chances, one will always end up winning. But all the chemical reactions that preceded this selection phase made use of both molecular alternatives with equal probability. There is no reason at all to assume that the multiphase process of the origin of life represents such a rare series of events that it was absolutely unique. But selection did act as a kind of filter that provided ultimately for the conservation of one continuous chain of events. It is through this historical chain that all life on earth can be traced back to one common source.

An a priori preference for left-handedness did not, then, account for the dissymmetry in the build-up of proteins, nor was it a unique chance occurrence that prevented symmetry in them. The annulling of parity as such was, despite the equal chances of winning that both competitors had, a necessary result of evolutionary behavior. This remains so no matter how often the so-called original event took place. We can ascribe the annulling of parity to a consistently competitive selection process based on non-linear growth rates (cf. Chapter 11). This annulling consequently represents an *a posteriori breach of symmetry* resulting from definite physical laws.

Our second example will illustrate a contrasting case, the *preference for a symmetrical solution.* The individual protein molecule has, as we have often mentioned, a definite structure and form but is by no means symmetrical. The only regular feature this molecule has is its peptide bond that joins the molecular building blocks together to form the macromolecular protein chain. The periodic repetition of this stereo-specific bond favors the formation of a helix with a definite direction of spiral. A normal protein molecule contains as many as several hundred chain links that can be classified under some twenty different types of protein building blocks. Each of these types is characterized by a special molecular side chain that represents a specific type of physical interaction. The sequence in which the individual elements of the chain appear reflects no symmetry whatsoever. Indeed, at first glance it appears to be completely arbitrary. But the entire secret of the functional efficiency of protein molecules lies in this clearly defined sequence.

In the course of a long evolutionary process, this chain that was originally formed by chance was gradually modified so that it could fold itself together in a characteristic way that permitted it to bring certain reactive groups closely together in the so-called active center of the molecule. These groups can now, like the individual parts of a machine, work together cooperatively on the "molecular assembly-line product," the substrate. We can regard proteins as the smallest machines known to us. They cut, weld, exchange, sort, transport, and transform, and each protein molecule or enzyme has its own specific task. Jacques Monod has therefore termed them "teleonomic structures." Their make-up conforms to no aesthetic principle. As in machines designed and built by human beings, functionality determines their construction. We will see, however, that functionality and symmetry can be synonymous. If we look at internal-combustion engines, we see that in most models the cylinders are arranged symmetrically in relationship to the crankshaft (see Figure 31). This arrangement allows the force from the pistons to be smoothly converted into the rotation of the crankshaft and also permits the opening and shutting of the valves to be precisely coordinated with the piston movements. In a similar way—or we might better say for similar "teleonomic reasons"—irregularly folded protein chains cluster in a mirror-image fashion. This allows them to match up their bonding capacities or to regulate their reactivity cooperatively.

The symmetrical protein complex that has been most thoroughly studied is hemoglobin, the substance that gives blood its color. It

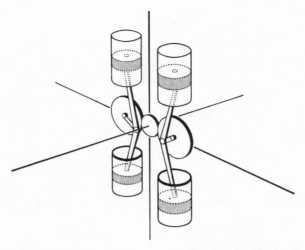

Figure 31. Schematic drawing of a four-cylinder internal-combustion engine with the cylinders placed in an "H" arrangement. The two crankshafts are joined by gears.

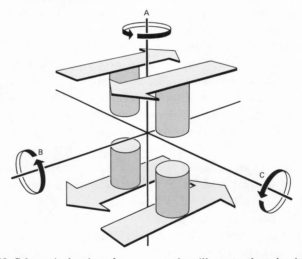

Figure 32. Schematic drawing of a hemoglobin molecule. Each of the four subunits (drawn as cylinders here) represents a protein chain whose structure is shown in Figure 11. The arrows indicate the orientation of the individual chains in the molecular complex. As the colors illustrate, the subunits make up identical pairs. Since the two sets of "twins" differ somewhat, the result is only a partial symmetry. Only axis A is an axis of a genuine symmetry. Axes B and C are axes of pseudosymmetries.[37]

consists of four molecular chains arranged in two pairs. These pairs are arranged in such a way that three axes, all perpendicular to each other, cross in the center of the molecule (see Figure 32). We owe our present knowledge of the exact spatial location of each of the approximately ten thousand atoms in this macromolecular complex primarily to the brilliant work of Max Perutz, John Kendrew, and of their Cambridge school. They have refined the method of X-ray structural analysis first developed by Max von Laue, William Henry Bragg, and William Lawrence Bragg to such an extent that we are now able to reconstruct precisely, from the reflections of X-rays, not only the iterative grid arrangement of the molecular crystal but also the by no means regular detailed atomic structure of the individual protein molecules.

Why do asymmetrical protein units come together in symmetrical fashion?

Even before the detailed structure of the hemoglobin molecule had been fully analyzed, Jacques Monod, Jeffries Wyman, and Jean Pierre Changeux had worked out a plausible model for it and thereby provided an answer to the question we have just posed. First of all, it is by no means impossible for asymmetrical elements to come together to form symmetrical figures. M. C. Escher has convincingly demonstrated this point in his drawings, which inspired us to attempt an illustration in his style of a "protein model." Figure 33 depicts two absolutely symmetrical figures, each composed of four fish. The fish in the first complex are huddled closely together. Their angular form suggests that they are hungry. In the second complex, the roundness of the fish indicates that they are well fed. A school of sardines had appeared on the scene, and one of the hungry fish immediately seized one. To maintain the symmetry, the remaining three fish were obliged to adjust their shapes to that of the first fish. In this larger and more open arrangement, it is easier for them to take nourishment and so to stabilize their round form.

The structural transformation of the four subordinate elements in the protein model developed by Monod, Wyman, and Changeux functions in an analogously cooperative way. It is the task of the hemoglobin molecule to bind oxygen and to carry it to the "metabolic

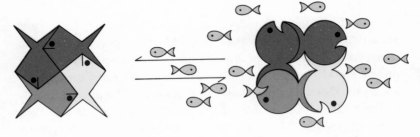

Figure 33. Allosteric model of a hemoglobin molecule. The transition between two alternative conformations is triggered by a substrate (in this case, the sardines) and proceeds cooperatively.

factories" in the body's cells. Each subordinate unit has an active bonding site; or, in other words, the molecular chains of hemoglobin are folded in such a way that centers for binding oxygen are formed. But the chains can also fold themselves differently and thus drastically reduce their affinity for oxygen. This happens as soon as they come into the presence of a "consumer" of oxygen. In the organism's metabolic centers, hemoglobin unloads all the oxygen it is carrying. In the lungs, however, where there is a large supply of oxygen, this transport molecule again assumes its high affinity for oxygen, just as our fish transform themselves once nourishment is present (cf. Figure 11).

This model of cooperative transformation of structure and of affinity regulation is by no means restricted to hemoglobin. Indeed, it describes even better the action of a "true" enzyme, which not only binds and transports a substrate, as hemoglobin does, but also catalytically transforms this substrate. The transformation of structure can be triggered here not only by the substrate itself but also by a specific activator molecule—such as the product of the reaction or some other by-product—that attaches itself to a different allosteric bonding site. We then have a regulatory mechanism that makes use of an activator to direct catalytic activity, initiating it or cutting it off as needed. In a similar fashion, several enzymes can be combined in a regulatory system, in which all individual functions are economically coordinated, just as they are in a well-organized chemical plant. The concept of allosteric reaction control has given us a profound insight into the mystery of coordinated reaction in the living cell.

But is symmetry an essential prerequisite for this process? Is there a natural law that contains a clause postulating the persistence of symmetry?

In answering this question we can again make use of our fish pattern. As soon as one fish changes shape, it no longer fits into the previous spatial arrangement. This is true to an even greater extent in the extremely complex patterns of bonding sites that establish contact between two protein chains. The forces activated by this temporary breach of symmetry tend to do one of two things. Either they force the subordinate unit in question to give up its substrate, to return to its previous condition, and thus to fit back into the original bonding pattern, or they so transform the other subordinate units that new contact areas adapted to the new form result. In our fish model, these alternatives are symbolized by the fact that all the partners either take on a round shape or retain the angular one.

But this does not fully answer our question of whether there is some mysterious law that works to preserve symmetry. After all, protein chains were not designed on a drawing board. In the course of evolution, two alternative structures for the same protein sequence must have developed, and they must have been so constituted that they both could be combined in a symmetrical manner to form a complex.

How was this difficult construction problem solved in an evolutionary process? How were the innumerable asymmetrical alternatives discarded?

In the folding of protein chains, those forms will always be favored that are optimally stabilized by the bonding affinity between the chain links; and in this respect, symmetrical patterns are by no means more advantageous than asymmetrical ones. A number of biochemists—among them Daniel Koshland—have shown that irregular enzyme complexes are just as able to regulate catalysis, as described above, as are their symmetrical counterparts. And if we return to our example of the internal-combustion engine, we can easily come up with a whole list of analogous asymmetrical designs. Modern diesel engines, for example, do not have the symmetrical structure of their predecessors, and the essential principle of the Wankel engine is the eccentricity of its rotor.

But then how can we explain that the postulate of functionality that is the sole determinant of evolution so often coincides with the demand for symmetry? It has been proved experimentally that as a rule enzyme complexes do have a symmetrical structure.

In evolution, only those mutants prevail that represent a selective advantage. In other words, if a mutant is to be selected, it has to be accompanied by a functional advantage that will in some way further the reproduction of that mutant. Among advantageous mutants, there are surely more that prefer asymmetrical to symmetrical patterns. But regular structures, wherever they occur, evolve more rapidly because in their case the advantage affects *all* subunits *simultaneously*. In asymmetrical structures, on the other hand, the advantage affects only *one* subunit, the one in which the mutation occurs.

This effect can be easily illustrated with bead games like the one shown in Figure 25. Certain configurations will be given higher selective value because symmetrical figures spread or reproduce themselves more rapidly. In a game without such rules, it would take a large number of comparable, though very different, mutations to arrive at the same goal. Since the evolutionary process has to pass through many phases before it arrives at a perfect final product, forms consisting of identical subunits will be characterized by a much higher rate of evolution. We find so many symmetrical structures in biology today because they were more efficient in exploiting their advantage and consequently won—a posteriori—the selection competition. They did not win because symmetry was—a priori—a necessary prerequisite for fulfilling their functions. Nature even tolerates a certain amount of deviation from perfect symmetry as long as functionality is not impaired. The symmetry of hemoglobin (Figure 32), for instance, is only a "near-symmetry." As Thomas Mann said, "Life shrinks back from absolute perfection."[33]

8

Metamorphoses
of Order

For the mathematician, the concept of order derives from the idea of definite ordering or arrangement. The physicist, on the other hand, sees "order" more as a contrast to "disorder." Ordered states of matter may include alternatives that are not comparable in any quantitative sense. The mathematician would correlate them in a scheme called a "partial order."

In the social life of human beings, the idea of a "just order" is central. Such an order is normative, not natural. The biological order has evolved primarily from the natural process of competitive behavior. If human beings want to establish norms for a "just order," they have to free themselves from this biological legacy; but in doing so, they have to retain their individuality as it is expressed in personally motivated actions.

The principles of thermodynamics determine the macroscopic, stationary behavior of inanimate matter. They also guide the impulse for order in the animate world, even though living beings are, in their individual structure, creations of chance.

Social thinkers often cite one or the other of these aspects of order to support their theories. But the invocation of "absolute and blind chance" as a scientific rationale for an "existential attitude toward life and society" is just as much an "animistic projection" as is the postulating of dialectical necessity as the basis for a materialistic world view.

8.1   A "JUST ORDER"

"The goal of social-democratic economic policy is a constantly rising standard of living and a just share in the fruits of our economy for all, a life lived in freedom, a life without degrading dependencies and exploitation."*

Who wouldn't enthusiastically support such a program? We would all rather be rich and healthy than poor and sick.

In his *Stories of Men and Numbers,* [38] Karl Menninger tells the parable of the two shepherds whose subjective views of justice severely tested their friendship. A stranger had given them eight dinars for sharing their evening meal with him. But the contributions they had made to the common meal were unequal. One shepherd had given five cheeses, the other only three. The first shepherd therefore demanded five of the eight dinars, but the other protested that this division was "unfair." The matter came before a cadi who asked the two shepherds to put friendship before the "naked law of numbers." But the two shepherds insisted on the "naked law." As far as they were concerned, there was only *one* absolute justice, and they expected the judge to reveal it in this case.

The judge arrived at the following verdict: The shepherd who had contributed five cheeses was to get seven dinars. The one who had given three cheeses was entitled to only one. The judge supported his decision with a simple calculation: Eight cheeses amounted to twenty-four thirds, and each man had eaten eight thirds. Since the first shepherd had contributed fifteen thirds and the second shepherd only nine, the stranger had eaten seven thirds from the first shepherd and

*Excerpt from the Godesberg Program of the Social-Democratic Party of Germany, November 1959.

only one from the second. The payment was therefore to be divided accordingly. The story goes on to say that the two friends, chagrined by the verdict, responded by giving the law of friendship precedence over the "naked law of numbers."

It is not our intention to oversimplify or make light of the highly complex problem of a just social order by relating this parable. All we mean to illustrate by it is that justice is normative and is not based on an absolute and predetermined order.

Before anyone attempts to replace a given order with a "more just" one, he should define as carefully as possible what is to be gained by the change. In this connection, we would like to quote a statement made in June 1969 by Karl-Dietrich Wolff, the national chairman of the Association of German Students for Socialism, in a conversation with the journalist Stephen Spender:[39]

> There are practically unlimited possibilities for utilizing the high productivity of our times, but all these possibilities are neglected because the system encourages waste, exploitation, and lust for profit, all of which work against the optimal utilization of these possibilities.

We offer as a response to this claim the following figures from the annual report on domestic business done by a leading chemical firm in the Federal Republic of Germany:

Gross Income: 10 billion DM	100%
Raw Materials and Supplies	47.6%
Depreciation and Overhead	21.3%
Wages, Salaries; Health Insurance, etc.	22.7%
Taxes (=60% of profits earned)	5.1%
Retained Earnings (added to the firm's reserve fund)	.7%
Dividends	2.6%

It should be stressed that 8.4 percent (the sum of the last three figures) represents a relatively good return, reflecting efficient management on the one hand and favorable economic conditions on the other. If we set these figures against the quotation that precedes them, we cannot but wonder what "practically unlimited" possibilities could be utilized if profit were abolished. When compared with the amount spent for wages and salaries, which keep increasing to meet the demands of the work force, the 2.6 percent allotted to dividends is minimal. And we should remember that this amount is divided among many people whose retirement income will be derived from the investment of their hard-earned savings in this firm. The broad distribution of this firm's stock, then, makes the firm an institution that serves the public interest. Abolishing profits would make many people poorer but none significantly richer. In industrial nations, a higher standard of living for all can be achieved for the most part only by the increased productivity of all and only in a few cases by a redistribution of capital.

Every political party today hopes to attract voters by claiming to favor "a just share in the fruits of our economy for all." But the question remains how, specifically and in detail, such a just distribution could be effected.

The first thing we should realize is that individual initiative will be increasingly limited as our sources of raw materials and energy become exhausted. (We will look into this problem in more detail in Chapter 13.) But it will still be possible to create wealth without taking anything away from someone else. If this were not so, there could be no "continually rising standard of living." What will always be needed, of course, is new ideas. Creativity will be just as crucial to our future as the ability to run the machines we invent. Realizing equality of opportunity cannot be achieved merely by unifying the school system. Increased educational opportunity is no doubt desirable, but we will derive little benefit from it unless we also develop specialized knowledge. There is much talk these days of "brainstorming" as a means for solving problems created by impending shortages of energy and raw materials. But without specialized education, we will lack the "brains" needed for "brainstorming."

In trying to assure everyone a just share in a rising standard of living—however we may choose to do this—we must not destroy the means we have for maintaining our standard of living. A high standard of living has to be earned by hard work. We cannot do without the incentives that produce that kind of work. Motivation presupposes rewards. We cannot simply discard this legacy of our evolution. If we want to establish norms of behavior, we cannot ignore human nature in doing so. Justice cannot be defined on an a priori basis; every society has to strive for it in terms of its own possibilities.

Economic growth rates will continue to decline (see Chapters 12 and 13). We will have to bring about a state of stability in such a way that the explosion in our growth rate does not suddenly give way to an implosion of decline. We cannot solve this problem by any kind of "leveling" or "equalizing" process. We are using this political concept intentionally because it has a negative connotation—the Russian word is *uravnilovka*—even in Marxist-Leninist usage. Lenin, and later Stalin, rejected equal pay for all as being un-Marxist. The only alternatives are state control of productivity and rewards for good work or individual initiative and responsibility in a free economy that admits reasonable "profit."

There is ample evidence, too, that shows how different social orders approach those other key points cited in our opening quotation, i.e., "a life lived in freedom, a life without degrading dependencies and exploitation." We find disagreement even on the definition of basic terms. Does "freedom" not mean "elbow room," room in which human capacities have free play, the freedom to move, to speak out, the opportunity to take chances, to hope, to believe, to love, and to be responsible for one's own fate? This is just as impossible in a totally regulated system that classifies all human effort in terms of "collective merit" as it is in a bacterial culture that grows unchecked.

The concept of solidarity originally seemed to express perfectly a need of our time. It emerged from the polarity between individuality and collectivity and took into account the interdependence of human beings. The individual's bond with society was balanced by society's respect for the individual. But the term solidarity has been eroded, if not totally corrupted, by constant uncritical use of it. Having been usurped by political parties, it is mainly used today to suppress unfavorable criticism. A call for solidarity accompanied by claims that criticism violates solidarity is a contradiction in itself.

Freedom needs order just as much as it needs room for innovation.

Since there is no mathematical formula for calculating justice in advance, justice can be realized only through an evolutionary process. In every phase of this evolution, the new must prove itself against the old. This is how life originated. This is how *Homo sapiens* developed. Only in this way can we achieve a life in freedom and put an end to degrading dependencies and exploitation.

8.2 THE ORDER OF NUMBERS

Mathematicians speak of an ordered set if all its elements can be compared according to definite criteria. The following rules must hold for any pair of elements a and b:

1. Either a is smaller than b, or b is smaller than a, or a equals b.
2. The three possibilities listed under 1 above are mutually exclusive. No two of these relations can exist at the same time.
3. Such a relation must be transitive. This means that if a is smaller than b and b is smaller than c, then a must be smaller than c.

This last condition reminds us of a story that Konrad Lorenz once told at the end of a lecture on pecking orders in the animal kingdom. In this particular case, the story had to do with wild geese, the favorite "pets" of the Max Planck Institute in Seewiesen.

A clear pecking order had established itself among three ganders we will call A, B, and C. The order was $A > B > C$. This means that whenever B met A, B made way for A. But if B met C, C showed B respect; and C naturally displayed great humility toward A. But, as was soon demonstrated, this order was not all that "natural." One day C pulled himself together and gave A a sound thrashing. From that moment on, their relationship was reversed, and for A, C was the "king of the pond." But this change affected only the relationship between A and C. Otherwise, nothing had changed. B continued to bow down to A, and C remained subservient to B.

This order can be abstractly represented by a cyclical pattern, in which the "greater than" symbol ($>$) expresses dominance:

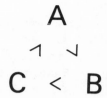

Similar cyclical hierarchies are common in sports. It happens quite frequently that team A will consistently defeat team B in a series of games, that B in turn will defeat C, but that C will usually win out against A. It is quite likely that psychological factors related to behavior in pecking orders come into play here.

In set theory, cyclical hierarchies of this kind would not be called orders or even "partial orders." Figure 34, illustrating the relationship between the three ganders, shows this clearly. The ganders cannot be depicted abstractly in a cyclical relationship unless each one is drawn twice. And in fact all the relationships in the cyclical hierarchy cannot be realized at the same time.

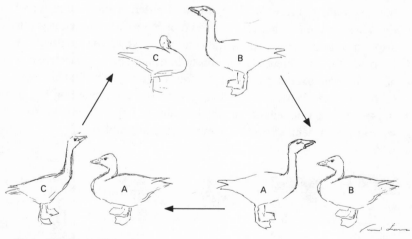

Figure 34. Scheme of a cyclical pecking order (sketch by Konrad Lorenz).

In physics, "partial orders" are very important. A set is called "partially ordered" when only the last two of the three conditions listed on page 136 are fulfilled. That means that not all the elements in a set can be directly compared with each other; there are alternatives that cannot be compared. But whenever a relation can be established, it must fulfill conditions 2 and 3, just as a complete order would. We are accustomed to dealing in such partial orders in our everyday lives. We may, for example, share with many other people a preference for certain pieces of music. But we would not be willing to set up a hierarchy that would place a partita by Bach above a string quartet by Beethoven, or vice versa. We feel that the individual beauty of both pieces does not permit of comparison and cannot be relativized by any value system.

The sequence of the natural numbers represents a precedent for a complete order. A special discipline in pure mathematics grew out of the study of these numbers, and the pioneers in this field of number theory were Pierre de Fermat, Leonhard Euler, Adrien Marié Legendre, and Carl Friedrich Gauss. We can use the subdivision of a natural number as an example of an order. Around the turn of this century the English mathematician Alfred Young introduced so-called partition diagrams to represent the relation of order. A diagram group of this kind for the number four is shown in Figure 35. These diagrams are defined as patterns of lines and columns in which the length of the horizontal lines decreases from top to bottom. The comparative size of two diagrams is determined in the following way: The first horizontal line in one diagram that is longer than the corresponding line in the second diagram makes the first diagram the larger. Using this rule, we can find a definite order for every natural number.

Figure 35. Young's order for the number four.

A few years ago, Ernst Ruch developed another definition of the inequality relation for Young diagrams.[40] In a fundamental study of the chirality problem (cf. p. 121), he used this definition to classify optically active chemical compounds.

Ruch's scheme can easily be laid out on a playing board. A horizontal row is filled with the number of beads corresponding to the number to be subdivided. The beads are then moved from the longer row to the neighboring shorter one, and the diagram that results each time is called smaller than the previous one. The only condition governing the moves is that no horizontal row can be longer than any one above it. If we have a set of more than five beads, alternatives result that can no longer be compared in terms of the inequality relation. A partial order has evolved from Young's order. Figure 36 compares the two diagram arrangements for the number ten.

Ruch's subdivision of numbers could be seen as a model for a "socialization of numbers." The numbers are subdivided in such a way that smaller subgroups increase at the expense of the larger ones. A particularly interesting point about this order is that it immediately produces variants which cannot be compared. In other words, if we progress in the diagram from top to bottom (this applies to numbers larger than five), we can do so by alternative routes; and in many cases, it will be impossible to change over to a different route without having to make a more or less major detour that runs counter to the direction defined by the inequality relation. Diagrams in different branches are therefore no longer directly comparable precisely because they lie on levels whose relation to each other is not clear. With larger numbers, the quantity of branches increases drastically. The Ruch scheme for the number one hundred is enormously complex, and for numbers relevant to molecular and physical phenomena (e.g., 10^{20}), it would assume dimensions beyond our grasp.

This concept is much more appropriate for representing physical phenomena than any scheme of perfect order, and its applicability to complex human and social relationships is even more obvious. We move toward a "just order" in society on a number of routes at once. Any claim that a single concept will bring salvation is clearly unrealistic in view of our historical experience.

n=10

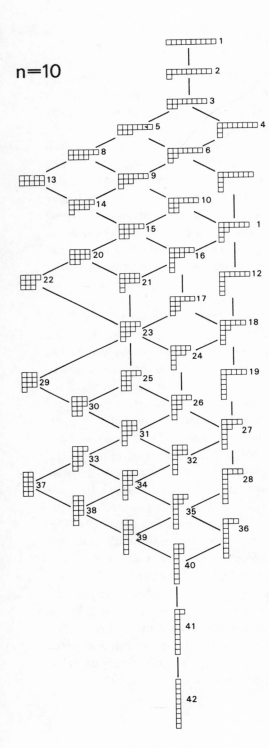

Figure 36. Ruch's partial order for the number ten. The numbers next to the individual diagrams indicate Young's order, in which all the diagrams are arranged in an unequivocal sequence.

8.3   THE ORDER OF MATTER

E N T R O P Y   In this section, we will give a somewhat detailed explanation of the extremely important concept of entropy, which, unfortunately, is not easily accessible to intuitive understanding. A grasp of this section is not essential for reading the chapters that follow. It does, however, enable the reader who understands it to have some idea of the application and the limits of this concept that is used so often in our age of statistics.

In our natural environment, we witness the spontaneous rise of order as well as the destruction of order. But physics is interested in natural processes only to the extent that they exemplify predictable and verifiable laws. Eugene Wigner stated this point succinctly: "Physics doesn't describe nature. Physics describes regularities among events and *only* regularities among events."

Physics must therefore exclude the complexity of natural constraints both in its experiments and in its abstract thinking. The physicist does this by isolating the systems he intends to observe from their natural environment or by subjecting the interplay between system and environment to rigorous controls. Johan Huizinga has remarked that every game creates its own special world. "It takes place within defined limits of time and space." This is also true of every scientific experiment.

The state of a closed physical system is defined by a number of parameters such as temperature, pressure, volume, and chemical composition. We have an intuitive understanding of most of these parameters. But our experience does not provide us with any means for grasping the most important variable affecting the organization of the physical world. This variable is *entropy.* Rudolf Clausius coined this word in 1865, deriving it from the Greek verb *entrepein,* "to turn" or "to reverse," and entropy does indeed have to do with reversibility, or, more precisely perhaps, with irreversibility. Clausius himself explained the choice of his term as follows:

In choosing an appropriate name for S (entropy), we note that the quantity U (internal energy) has been defined as the heat and work content of a body. By analogy, we could describe the quantity S as a body's transformational content. I think it a good idea to give such important scientific concepts names derived from the ancient languages so that any modern language can adopt them unchanged. I would suggest calling the quantity S the "entropy" of a body. This term is based on the Greek word *tropae*, meaning "transformation." I have deliberately made the structure of this word analogous to that of "energy," because the two quantities described by these terms are so closely related in physics that the parallel seems useful to me here.

Although the term entropy has now been in use for over a hundred years, those not trained in physics continue to approach it cautiously and respectfully. Encyclopedia articles have not done much to improve this state of affairs. Simultaneous references to "caloric state variable," "measure of probability," "parameter of order," or even "parameter of information" have contributed to the general confusion and led readers either to underestimate the term's significance or to overestimate grossly its interpretive value.

Entropy is a parameter that can be formulated quantitatively, and that can be used in characterizing both the distribution of energy among the various quantum states of a physical system and the informational content of an encoded message.

It might be useful at this point to understand historically why the introduction of this concept became necessary. To do this, we shall have to assume the perspective of physicists and chemists in the first half of the nineteenth century.

Newton's concepts of absolute time and absolute space, as well as his linking of these concepts in the laws of mechanics, had produced a view of the physical world as a closed system dominated by cause and effect. At the same time, chemists were constantly producing new evidence to substantiate Dalton's hypothesis that the atomic structure of matter was based on definite proportions. It seemed only natural to apply Newton's mechanics to the smallest units of matter in an effort to arrive at a unified description of all the characteristics of matter.

An apparently insurmountable problem, however, stood in the way of this effort. In order to solve Newton's equations, scientists needed to know the basic data for *every* particle of matter, i.e., they had to know the spatial coordinates and velocities of any moving particle. But this was clearly a hopeless task, since just a single drop of water contains approximately 10^{21} (a trillion billion) molecules.

As a rule, scientific problems cannot be solved merely by brainstorming. But sometimes they can be circumvented, and that is what was done in this case. The attempt to describe the course of all individual processes in deterministic terms was given up completely.

In the late eighteenth and early nineteenth centuries, advances in mathematical statistics and in probability theory helped create the insurance business as we know it today. It was possible, for example, in dealing with large numbers and without having any information about individual cases, to make fairly accurate predictions about average life-span or the frequency with which certain accidents and natural disasters occur. It was also known that the larger the sample was, the more accurate a prediction based on the laws of statistics would be.

That matter behaved in an analogous fashion seemed likely. Properties like temperature, pressure, etc., had to represent in some way the average behavior of individual particles. By applying the laws of statistics, it should be possible to correlate these properties exactly with the average mechanical properties of individual particles. The Englishman James Clerk Maxwell and the Austrian Ludwig Boltzmann followed these ideas to their logical conclusion and thereby opened up new scientific vistas. But Maxwell and Boltzmann did not undermine the foundations of physics by applying the statistical concept to it. Philosophers could continue to apply the principle of causality in their dialectics. It was not until the twentieth century that physicists, drawing on the insight that statistical behavior may be inherent in the elementary processes themselves, formulated a concept central to modern physics: the uncertainty principle.

The immediate consequence of the statistical way of thinking was the positing of entropy as a new variable of state. Seen historically, the formulation of entropy as a variable was closely linked to the definition of temperature. Temperature can be correlated with an average property of individual particles. For an ideal gas, the meaning of this average is particularly clear and simple: Temperature is directly proportional to the average kinetic energy of the particles. It represents an "intensity" of behavior and is therefore called an "intensive" quantity. As such, it is not dependent on the number of particles from which the average is derived. But on the other hand, it follows that there must also be an "extensive" quantity that is complementary to temperature and reflects an amount or "extent" related to this intensity. Without such a complementary quantity, information about the thermal energy of the total ensemble of particles in question would be incomplete.

"Information" is the key word here. Entropy, which is complementary to temperature, is a measure of information. It tells us how the total energy is distributed among the various quantum states of a system.

Whenever we represent a system by an average, we necessarily lose some information about the total system. Consequently, we have to specify how many individual cases are included in taking that particular average. This is a measure of the information that was lost by simply specifying the average.

The following example illustrates this principle. If an airline wants to determine how much weight a jetliner is carrying, it does not ask all the passengers to step on a scale. For domestic flights within the United States, not even the total weight of the luggage is checked. The airline simply limits the pieces of luggage allowed per passenger. It assumes an average weight for each passenger, i.e., a preestablished intensive quantity. Given this quantity, all that then matters is the number of passengers. This extensive quantity alone supplies all the specific information needed to determine the actual load. This example clearly illustrates distribution in terms of intensive and extensive variables. The intensive quantity is independent of the number of passengers and has, for any type of cargo, a preestablished average value.

But calculating entropy in physics is not as simple a matter as determining the weight of cargo for airplanes. A physicist would immediately object that our example has neglected a crucial point: Energy can be exchanged between particles; airline passengers cannot exchange their weight, much as some of them might wish to trade figures with a trim stewardess. The airline simply acts as if every passenger had the same average weight, an assumption that is justified only if the number of passengers is large enough.

Calculating averages for entropy is considerably more involved, and cannot be achieved merely by counting quantum states or particles of matter. Temperature can be established only if an exchange of energy between particles actually occurs and if a thermal balance can then establish itself. This can happen in a closed system provided the total energy and the total number of particles remain constant. The physicist first has to find out how an individual molecule absorbs energy and distributes it among the various degrees of freedom related to translation, rotation, and internal vibration. (Atoms and molecules can toss energy back and forth to each other only in "balls" of "quantized" size.) Working on the basis of the exchangeability of energy quanta, the physicist must then find the average of all possible combinations that are compatible with the total energy, which, in an isolated system, is a strictly conserved property. This task of averaging was somewhat simplified by quantum theory, which posits that similar particles at the same energy level are indistinguishable. However, the number of atomic and molecular states—comparable to the individual distributions of beads we have already seen in statistical bead games—is in any given case immeasurably large in relation to the total number of states in which the absorption of energy can occur. These states correspond to the number of squares on the board of our bead games.

An example will help to make these somewhat abstract considerations concrete and will also show that entropy is a general measuring tool for characterizing statistical distributions. As such, it is by no means applicable only to problems of thermodynamics. Let us consider a printed passage consisting of one hundred symbols. By symbols, we mean the twenty-six letters of the alphabet plus three punctuation marks and the space between words. We are using, then, a total of thirty different symbols.

How many different passages of one hundred symbols could be produced?

If we had no knowledge of the structure of English, and if every conceivable combination were admitted as meaningful, then there would be exactly

$$30 \times 30 \times 30 \ldots \times 30 = 30^{100} \approx 10^{148}$$

possible passages. Each of the hundred positions could theoretically be filled with any one of the thirty symbols. The total number of possible combinations or microstates is thus 10^{148}. If we wanted to arrive at a given passage simply by guessing and not apply any system whatsoever, we might have to try 10^{148} times. We could also express this by saying that the probability of arriving at a given sequence of symbols on the first try is only one in 10^{148}.

If we lengthened the passage to include two hundred symbols, we would then have $10^{148} \times 10^{148} = 10^{296}$ possible combinations. If the passage is doubled in length, the number of possible combinations is the square of the original number. If the passage is made ten times as long as it originally was, the possible combinations increase to the original figure to the tenth power. Entropy, as an extensive quantity, should represent a property of size. If the size is doubled, then entropy too should be doubled. Therefore entropy should not be represented directly by the number of possible combinations or microstates. It is rather the logarithm of that number that correctly reflects this relationship. Logarithms convert a measure of probability manipulated by means of multiplication into a measure of quantity manipulated by addition: $\log [a \times b] = \log a + \log b$.

In short, entropy is represented by the logarithm of the number of possible combinations. This quantity is in itself without dimension. In thermodynamics, multiplication by the Boltzmann constant (named after the founder of statistical mechanics) allows us to assign the product of temperature and entropy the dimension of an energy. The reason for thus providing entropy with a "caloric" dimension is purely historical. It would have been just as possible, by multiplying temperature by the Boltzmann constant, to assign temperature alone the dimension of an energy. In that case, entropy, like its corresponding quantity in information theory, would be without dimension. For practical reasons dictated by the logic of decision-making in computers, information theory uses a logarithm to the base 2 and expresses this by speaking of "bits," i.e., binary digits. *Entropy is then merely the number of binary decisions needed to identify a sequence of symbols.*

In a distribution of energy, different quantum states have different "weights," just as in language certain letter combinations among all those possible occur with much greater frequency than others.

In language, this is essentially the result of two factors: (1) not all symbols occur with the same a priori probability, and (2) not all letter combinations that conform to a probability distribution make up meaningful words, much less sentences.

In our first estimate, we regarded all conceivable combinations of letters as equally probable and therefore as equal in rank. In all languages, the space between words is the symbol that occurs most often. But a sentence that was made up entirely of this symbol—in other words, an empty line—would be meaningless. If we take into account the empirically established frequency with which letters occur—in English, the most frequent symbol, apart from the space between words, is *e;* the least frequent is *x*—entropy is reduced significantly, in English by about 16 percent. Thus, instead of $\log_2 30$ or 4.91 bits, we need on the average only 4.12 bits* to guess one letter systematically. The number of possible combinations (microstates) therefore decreases by a factor of *ca.* 10^{24} from 10^{148} to about 10^{124}.

The reduction of entropy can be easily simulated in a bead game. It is best to use a large playing board on which each square is assigned a letter. Letters will be represented on the board in proportion to the occurrence in the language. If we now try to form meaningful words out of the letters selected by the dice, this will be easier to achieve if

---

*It would take exactly five binary decisions to guess one symbol out of thirty-two equal classes of symbols ($2^5 = 32$). To be sure of making that number of decisions, we would, of course, have to proceed systematically. We could best do this—just as the computer does it—by representing each symbol by a sequence of five binary digits (e.g., 01001) and by running through the digits in order.

we start out with this realistic distribution of letters rather than with a homogeneous one. We will return to games of this kind in our discussion of information and language contained in the last part of this book (see Figure 63).

The second factor mentioned above, the one that takes the meaning of letter combinations into account, affects the reduction of entropy much more than does the first factor, which reflects the differing frequencies with which letters occur.

Claude Shannon, the founder of information theory, developed a game to demonstrate this phenomenon empirically. [41, 42] The first player thinks up a sentence. The second player then has to guess this sentence letter by letter by asking only yes-or-no questions. The number of questions he needs is recorded and subsequently compared with the number of questions that would be necessary if one had no knowledge of the language at all. It is then possible to calculate from this comparison by how many units the bit number of 4.91—the one resulting from a homogeneous and random distribution of letters— would be reduced by a knowledge of the language, of its structures and redundancies. Shannon analyzed a hundred texts and found an entropy reduction ranging from 50 to 70 percent, i.e., from 2.5 to 3.5 bits, per symbol. We should note, of course, that the players were students of information theory and were familiar with the frequencies with which letters occur, with average word lengths, and with sentence structure. Needless to say, completely unfamiliar texts were used. As we would expect, the players exhibited the greatest uncertainty when beginning words and sentences, but were able to complete them with relative ease.

What this shows is that different microstates have completely different statistical weight. Most sequences of letters can be disregarded from the outset because they do not make any sense. This principle also applies to the distribution of energy among quantum states. There is a characteristic probability for any individual quantum level, just as there is for any combination of symbols. In our example of a sentence with one hundred slots occupied by arbitrarily distributed symbols of thirty equal classes, the probability for any one possible combination was homogeneously one out of 10^{148}. However, if we have a meaningful sentence in a known language, the probability for many letter combinations will be zero, but for others it will be correspondingly higher than one out of 10^{148}. If we are to be able to compare different sentences with each other, the probabilities have to be normalized or calculated in terms of their "share" of probability. In other words, the sum of probabilities for all possible combinations must always equal one. Entropy then represents the average value of the (negative) logarithm of these probabilities. It should be noted that the logarithm of a number smaller than one is negative ($-\log_{10} \frac{1}{10}^{148} = +148$). We arrive at the average value by multiplying each logarithmic term by its probability and by summing all terms.

This expression for entropy formulated by Boltzmann and Shannon is used today in many ways and, unfortunately, often uncritically. We hope to have shown that its use is appropriate only in cases where average values convey meaningful information. This is clearly the case in speaking of the distribution of energy among the different quantum levels of a molecular system. Here the individual distribution is of little interest. The chemist wants to know, for instance, how much heat is produced in a reaction and what temperature is optimal for a technological process. A knowledge of entropy is equally useful in various aspects of communication science and technology.

It is important to determine carefully in each individual case whether the crucial elements in a problem are not lost in the process of averaging. For instance, Ruch's partition diagrams (see page 140) provide much more detailed statements about a distribution than the mere average value of entropy can supply, although they still involve averages. They offer heretofore unexploited possibilities for expanding statistical theory so that it can reflect details more accurately.

On the other hand, we are doubtful about the application of the entropy concept to so-called information aesthetics. Our criticism is not aimed at the use of information theory in the study of aesthetic problems but at the application of procedures based on average values to cases where the crucial information is contained exclusively in details.

Applications of the information concept in aesthetics generally result in comparative statements. Different concert programs have been compared in terms of their "originality," e.g., by Abraham A. Moles.[43] In these attempts, normalized probabilities have been used in calculating entropy values. This makes any comparison meaningless (cf. p. 308). The confusion that can result if normalization is left unclear in the subdivision of a set can be illustrated in the form of a story* that may provide the reader with some relief after the difficulties of this chapter.

Ali Baba had four sons. When he died, he left thirty-nine camels to his sons, and he had provided that his legacy be divided among his four sons as follows: The oldest son was to receive half of Ali Baba's property; the second, a quarter; the third, an eighth; and the youngest, a tenth.

The four brothers were at a loss as to how they should divide the inheritance until a stranger came riding along on his camel. The stranger knew just what to do.

He added his own camel to Ali Baba's thirty-nine and then divided the forty among the sons. The oldest son received twenty; the second, ten; the third, five; and the youngest, four. One camel was left. The stranger mounted it—for it was his own—and rode off.

Amazed, the four brothers watched him ride away. The oldest brother was the first to start calculating. Had his father not willed half of the camels to him? Twenty camels are obviously more than half of thirty-nine. One of the four sons must have received less than his due. But figure as they would, each found that he had more than his share. They decided to consult a sage and ask him to explain this miracle.

The sage calculated that if they divided thirty-nine according to the formula Ali Baba had prescribed, there would always be a remainder of

$$39 - 39(\frac{1}{2} + \frac{1}{4} + \frac{1}{8} + \frac{1}{10}) = \frac{39}{40}.$$

But since Ali Baba had provided that his entire legacy was to be divided among his sons, the sons would have to apply the same

*An abridged adaptation of the title story from Karl Menninger's *Ali Baba und die 39 Kamele (Ali Baba and the Thirty-Nine Camels)* (Vandenhoeck and Ruprecht, Göttingen, 1964).

formula to any remaining part of the inheritance. In that way, they would come to the correct result in the end.

It is probably safe to assume that the four brothers did not grasp the significance of a geometric progression and its convergence. But they had unlimited confidence in the sage and in his mathematical abilities, and they let themselves be convinced that everything was in order.

The sage could, of course, have told the brothers right away that the sum of ½, ¼, ⅛, and ¹⁄₁₀ is not one, and that by normalizing the sum of the denominators they could instantly have solved the problem of dividing thirty-nine without a remainder. All they needed to do was divide each share by the sum of the fractions (or multiply by ⁴⁰⁄₃₉).

EQUILIBRIUM   Rudolf Clausius concludes his classic work on entropy with these words: "The energy of the world is constant. The entropy of the world strives toward a maximum."

These days we are somewhat more cautious in our statements. John Archibald Wheeler summarized the modern view of order in the universe in this way: "We can believe that we will first understand how simple the universe is when we recognize how strange it is."

Thermodynamics would not be the marvelous and logically consistent discipline it is if it had indulged in speculations on the universe. The strength of its argumentation derives from the very fact that it deals with isolated systems whose initial conditions and boundary values can be controlled and reproduced. Thermodynamics is therefore a science that is more readily applicable to systems that can be studied in the laboratory than to the universe. As such, it provides insights primarily into systems, both animate and inanimate, with defined boundaries.

Clausius's two statements on the energy and entropy of the world, when applied to closed systems, form the basic axioms of thermodynamics.

The first law of thermodynamics articulates a principle of invariance or conservation, and represents one of the basic symmetry relations we have often referred to in this book.

The second stands for the asymmetry in natural processes. Indeed, it contains the only basic natural principle that includes a preferred direction in time. This seems paradoxical at first, because the laws of mechanics, including those of relativity theory and quantum mechanics, do not specify a direction in time (cf. p. 115). How then can it be that such an asymmetry appears in entropy, the statistical foundation of which is based on the laws of mechanics? Should not the reversal of a mechanical process, the direction of which is not determined a priori, produce a reversal of sign in the direction of entropy? But this in turn would contradict the second law of thermodynamics.

This objection to a statistical and mechanical interpretation of the second law of thermodynamics is almost as old as the theory of thermodynamics itself. It was first raised by Ludwig Boltzmann's contemporary and countryman Joseph Loschmidt. The elucidation of this point is of the greatest importance if we want to understand what forces provide order in the animate realm.

The second law of thermodynamics says that the entropy of a closed system will continue to increase until the system achieves equilibrium. The "internal" production of entropy—that is, the entropy produced per unit of time by the processes at work in a system —is always positive and does not fall to zero until equilibrium is reached. If a constant temperature is maintained, this principle also applies to systems that are not closed, that is, systems that carry on heat exchange with their environments. What is important here is to differentiate between the "internal" production of entropy and the flow of entropy connected with heat exchange. The state of equilibrium itself is a stable state that can be altered only by external physical changes. All that internal fluctuations can bring about are local (microscopic) reductions of entropy that will, however, in accordance with the second law of thermodynamics, always automatically balance themselves out. The Ehrenfest game (cf. p. 35) provides a model for a typical case of such equilibration, including microscopic fluctuations.

Everyone intuitively understands the nature of equilibrium. A balance scale is in equilibrium when both its pans hold equal weights. The arc of a pendulum is centered on its axis of equilibrium. Equilibrium means a "balance of power" in mechanics and thermodynamics as well as in politics. Arthur Schopenhauer found an apt image for it:[44]

> On a cold winter's day, a group of porcupines huddled together to stay warm and keep from freezing. But soon they felt each other's quills and moved apart. When the need for warmth brought them closer together again, their quills again forced them apart. They were driven back and forth at the mercy of their discomforts until they found the distance from each other that provided both a maximum of warmth and a minimum of pain.
>
> In human beings, the emptiness and monotony of the isolated self produces a need for society. This brings people together, but their many offensive qualities and intolerable faults drive them apart again. The optimum distance that they finally find and that permits them to coexist is embodied in politeness and good manners. The English warn anyone who comes too close to keep his distance. Because of this distance between us, we can only partially satisfy our need for warmth, but at the same time, we are spared the stab of each other's quills.

A system on which forces are at work will keep changing until it reaches equilibrium, i.e., until it is, on the average, free of such forces. This fixed state in no way implies that the system is "frozen." Both energy and matter can continue to flow back and forth between its constituent parts. In a similar fashion, a balance scale whose trays are equally weighted can continue to teeter back and forth. Indeed, we can get a more precise reading when the pointer on the scale is still in motion.

In the example of the scale, equilibrium is true equilibrium whether we introduce energy into the system and set it in motion or not. The symmetry of equilibrium will be reflected in the symmetry of the oscillations. If the beam of the scale is moved, the scale takes on potential energy that is proportional to the square of the distance of deflection and is independent of the direction of motion. If the system is then left alone to oscillate, potential energy will constantly be transformed into kinetic energy (and vice versa), but the total energy, apart from losses through friction, remains constant. The kinetic energy is proportional to the square of the velocity of motion at any time. The quadratic relation applying to both types of energy directly expresses the symmetry of the state of equilibrium.

Because of friction, the oscillations will gradually slow down and eventually cease. In a thermally isolated system, the total energy of the correlated oscillations would be transformed into heat energy represented by the uncorrelated motion of the atomic building blocks of the material the scale is made of.

We will now take a brief look at this special form of energy. In a closed system (isolated against any form of heat exchange) the internal energy is constant. Here, too, equilibrium means the distribution of energy to all available quantum levels. The state of the system can be characterized by thermodynamic parameters like temperature, entropy, pressure, volume, and (relative) proportions of the chemical composition. These parameters are constantly subject to spatial and temporal fluctuations, but in contrast to the equilibrium model of the scales, their changes never build up to the level of independent periodic motion, not even as a consequence of an externally induced disturbance. Because of the constant exchange of energy and momentum among all the molecular and atomic particles, a macroscopically correlated acceleration never occurs. Lars Onsager[45] first formulated a principle that accounts for this phenomenon.

There is, however, a striking parallel between these two types of equilibrium (i.e., mechanical and thermodynamic), and this parallel is rooted in a symmetry relation. In a basic work on the thermodynamics of decay processes near equilibrium—these are known as relaxation processes—Josef Meixner[46] called attention to this analogy. As in the example of the scale, internal energy and entropy can be represented as quadratic forms of "perturbation parameters," which are, in this case, the minute deviations of thermodynamic variables from their equilibrium values.

The Ehrenfest urn game (see Table 3) demonstrates this relationship in a particularly vivid way. Equilibrium is expressed in the equal distribution of different colored beads. On a playing board with 64 squares, for instance, there will be 32 black and 32 white beads. Any deviation from these expected values can be called a "perturbation parameter." As Figure 6 shows (and we can verify this ourselves by playing the game), the probability of deviation from the average value follows a Gaussian bell-shaped curve that displays a quadratic shape near the equilibrium point. This quadratic shape is also characteristic of changes in entropy, and it expresses the symmetry of the system around its equilibrium point.

Any possible deviation of entropy from equilibrium value has a negative sign. Entropy can only *increase* as equilibrium is reestablished, regardless of the direction the shift of the chemical equilibrium has taken. Near the point of equilibrium, then, the distinction between past and future cannot be derived from the behavior of entropy. Our awareness of the passage of time is due to the fact that the realm of the universe in which we live is still nowhere near a state of equilibrium. The overall equilibration process is not far advanced, or, as Friedrich Hund expressed it, our world is still "relatively young."[47]

The bead game described in Table 3 shows that true temporal asymmetries result solely from the initial conditions of the bead distribution, not from "natural laws," or the rules of the game. The statistical formulation of the second law of thermodynamics states, among other things, that every specific pattern of a distribution will dissipate in the course of time. As we can see by counting the number of possible distributions, the state of equilibrium can be realized in a maximum number of ways. Any specific individual arrangement of beads (one defined by a specific pattern of coordinates) with the same number of black and white beads is just as improbable as one of the two extreme asymmetrical distributions with all black or all white beads.

This makes Loschmidt's objection about time reversal immaterial. Time reversal means the projection of an experienced past into an uncertain future. It implies the pinning down of a pattern that is bound to dissipate like any other specific pattern. Ilya Prigogine[48] has provided entropy with a statistical formulation that explicitly takes this point into account and thereby emphasizes the true character of entropy as a measure of information, as a measure of our knowledge about reality. In Prigogine's interpretation, time reversal in a state of non-equilibrium therefore represents a negative jump in entropy, accompanied by a compensation. This compensation reflects the fact that the projection of the past into the future is a dissipation of present information, that is, a dissipation of the pattern established at the moment of time reversal.

We are able today to simulate these extremely difficult processes in experiments. In connection with the parity problem (cf. Figure 29), we mentioned the possibility of influencing the rotation of atomic nuclei by placing them in a magnetic field. Let us assume we have a collection of such atomic "tops" that are capable of exchanging energy among themselves in such a way that, after an equilibrium has been established, we can speak of a rotational or spin temperature. By manipulating the surrounding magnetic field in certain ways—i.e., by means of some "external" interference—we can simulate something like a time reversal. Since a certain amount of time is needed to bring about the equilibrium that determines spin temperature, we can register the echo of the process that immediately preceded the moment of time reversal. The results from experiments of this kind, such as those conducted by John S. Waugh[49] and his colleagues, are completely consistent with what statistical theory would lead us to expect.

We began our discussion of thermodynamic equilibrium with a quotation from Rudolf Clausius. How do his two statements hold up in the light of current knowledge?

Clausius's statement about energy, if we expand it in the light of relativity theory and include in it the equivalence of mass and energy, belongs to those basic axioms of physics whose validity experience has demonstrated beyond any doubt. His statement about entropy, applied to that part of the world accessible to us, suggests that we are still in a very early stage of development, that is, that we are far from reaching equilibrium. Our consciousness of time, too, oriented to mutability and inherent in the life rhythm of our brain cells, is rooted in the irreversibility posited in the second law of thermodynamics.

We are still in total ignorance of what consequences the statement on entropy may have for the world in the distant future. The future lies in darkness, in the darkness of the night sky, we might say. It gets dark at night because all the light that we receive from the sun—apart from a small amount reflected by the moon and some planets—has to come to us directly. It is not reflected by the cosmos; the universe is not in an equilibrium of radiation. Physicists have been studying this phenomenon for some time; it is caused by the expansion of space, an expansion that takes place with the speed of light.

Even in our solar system thermodynamic equilibrium cannot be established unless the sources of energy are exhausted. If irreversibility ceased to exist, we would never know it, for the very asymmetry of unachieved equilibrium is one of the essential premises of all life. The disappearance of this asymmetry would mean "heat death" for the entire universe and therefore the end of all life.

What would follow?

Timelessness?

Or time reversal concomitant with a contraction of the universe?

With these questions we have far overstepped the limits of present knowledge.

8.4   THE ORDER OF THE LIVING

CHAINED DEMONS   The appearance of our world reflects order rather than disorder. The second law of thermodynamics has often been interpreted as though it worked against any striving for order. So expressed, this interpretation is incorrect. Given the conditions we experience in our environment, we might even say that thermodynamic equilibrium represents a high degree of order, not only in terms of definite proportions of substances but also in terms of their spatial arrangement. As Carl Friedrich von Weizsäcker once put it, "Heat death consists not of mush but of many skeletons."[50]

The order we encounter everywhere around us can be explained by the fact that the temperatures on our planet are, in Erwin Schrödinger's phrase,[51] "in the neighborhood" of absolute zero. Compared with the temperature of the sun, which is between 4,000 and 7,000 K (K = one degree on the Kelvin scale) on the surface and between 17 and 21 million K in the interior, temperatures around 300 K that we experience on earth are indeed "in the neighborhood" of absolute zero.

The development of order and equilibrium does in fact conform to the postulates of the second law of thermodynamics, that is, to an always positive, "internal" production of entropy. This production of entropy is fed either by internal sources such as the energy localized in atomic and molecular interactions or by flows of heat into the system. There are no other possible ways for order to arise spontaneously in a closed physical system. Otherwise, "demonic forces" would have to be at work.

Hypothetical "demons" have repeatedly been invoked in attempts to provide a plausible explanation for Clausius's entropy law. Léon Brillouin has summarized these attempts in the following proposition: Even demons need a metabolism to do their work, and the concomitant production of entropy would make up for any deficit in the entropy balance.

We shall now take a closer look at these demons, which were, of course, postulated only to explain certain paradoxes and not because their "inventors" really believed in them. This will bring us directly face to face with the problem of biological order.

1. *Maxwell's Demon.* In 1871, James Clerk Maxwell described in his theory of heat ". . . a being whose faculties are so sharpened that he can follow every molecule in his course, and would be able to do what is at present impossible to us. . . ." This being can open a flap in a partition that separates a box into two halves. Whenever a fast molecule approaches the partition from the right half of the box, the demon opens the flap and lets it through. At the same time, he takes care to keep the flap closed to slow molecules. On the left side of the box, he does just the opposite. He lets only the slow molecules into the right half and keeps the flap closed to the fast ones. In this way, all the fast molecules gather in the left side of the box and all the slow ones in the right. The temperature in the left half of the box climbs while that in the right half drops.

We can build a device that accomplishes this task. A gas is compressed, and after a heat exchange with the environment has taken place, it is transferred into another container where it expands adiabatically (i.e., without exchanging heat with its environment). Since the heat created in the compression chamber was conducted off, a cooling process now takes place in the expansion chamber. What is important here is that there is *no microscopic mechanism* that separates the slow and fast molecules and thus—in a closed system with a homogeneous temperature distribution—

spontaneously creates a gradient. A process like this is possible only by external manipulation, that is, with the help of a machine that uses up energy. As every refrigerator owner knows, "cold calories" are even more expensive than "hot calories."

2. *Loschmidt's Demon.* Here we have to imagine an elf that can activate a time switch that suddenly reverses the flow of time and projects all the processes of the past into the future. We have already discussed the consequences of such a time reversal in detail in the preceding section.

Ilya Prigogine has shown that the gain of information produced by such a reversal is equivalent to a leap in entropy and therefore has to be "paid for." Consequently, this demon, too, is "chained" by the second law of thermodynamics. Once again, as we showed with the magnetic echo experiment described on page 156, we can construct a machine that can simulate—with a considerable cost in energy—the effect of a time reversal.

3. *Monod's Demon.* In his book *Chance and Necessity,* [52] Jacques Monod writes:

The second law of thermodynamics makes only a statistical prediction and therefore does not exclude the possibility that any given macroscopic system, in the course of a transformation of minimal range and for a very short period of time, could slide back down the "slope" of entropy, i.e., somehow move backwards in time. In living beings, it is precisely those rare and ephemeral transformations that are preserved by selection once the replication mechanism has picked them up and reproduced them. Selective evolution depends on the choice of those few and valuable mutations that turn up among innumerable others that macroscopic chance produces. In this sense, selective evolution is a kind of machine that permits us to move backwards in time.

In another passage, Monod indicates what he imagines this "machine" to be: "Enzymes function, after all, just like Maxwell's demon after Szilard and Brillouin corrected our conception of it: They tap chemical potential by way of routes laid down in a program that they carry out."

Prigogine[48] vehemently opposes such an interpretation, since that hypothetical "rectifier" of entropy fluctuations could not possibly exist.

He suggests instead that these arguments, derived from the statistical mechanics of equilibrium processes, simply do not apply to situations far removed from equilibrium. He concludes: "Far from being the work of some army of Maxwell's demons, life appears as following the laws of physics appropriate to specific kinetic schemes and to far-from-equilibrium conditions." We assume that Monod—like Maxwell and Loschmidt before him—was aware that such "demonic" forces in microscopic form cannot be active in equilibrium systems. This is why he speaks of a "machine." In no case is it the enzymes that can act as a microscopic fluctuation rectifier by virtue of their construction. In equilibrium, enzymes too work with complete reversibility.

An analysis of corresponding non-equilibrium situations shows that, even far from equilibrium, physical laws make impossible a spontaneous and continuous rectification of fluctuations, a task we might expect a demon (read: a microscopic machine) to perform. Here, too, ordering must take place by means of an energy flow that is maintained macroscopically. This was also the case in any mechanical realization of the effects produced by Maxwell's and Loschmidt's demons.

We will try to clarify this issue by studying a system that is far removed from equilibrium. We are less interested in the establishment of thermal equilibrium, which may occur within a millionth to a billionth of a second, than we are in a chemical transformation that takes place at a finite, easily measurable speed. As long as the system is not in equilibrium, forces are at work to bring it into equilibrium. In this process, "detours" are permitted, provided the system is still nowhere near equilibrium. The system can be prevented from reaching equilibrium by constantly introducing fresh reactants and by constantly removing the reaction products. Indeed, the system can be so regulated that there is a constant ratio of reactants to products. It then appears as if the system were in equilibrium because the ratio is constant and does not change over time. But in contrast to true equilibrium, this situation continually produces entropy. There will, of course, also be statistical fluctuations, and we will have to see what effects these have on the stability of the stationary state.

We have already conducted a detailed analysis of this kind in Chapter 4 with the help of statistical bead games. There we found three basic possibilities. The stationary state of reference is stable—as it always is in genuine equilibrium; or it is indifferent, that is, the ratios do not remain constant but change continuously; or the situation is unstable. In the latter case, the fluctuations deviate so far from the reference state that they reach macroscopic dimensions and lead to a complete collapse of that state. The stable state is the only one based on premises applicable to equilibrium as well (cf. p. 154). Of course, one important secondary premise is not fulfilled: The stationary state results solely from the compensating effects of formation and decomposition rates—an example is formation by autocatalytic reaction accompanied by the removal of wastes from the system—not from true reversibility, i.e., from a reciprocal reversal of the two processes. This means that the premises for one of the most important symmetries of equilibrium, the premises that postulate absolute microscopic reversibility for every step of the process, do not apply.

The difference between stability and instability can be illustrated by the example of a vote of no confidence and its consequences in a parliament. The government party may manage to garner enough votes in its favor to stay in power. In this case, the composition of the parliament will change little, if at all. In other words, the situation can be stabilized. Or the opposition may receive the majority of the votes. In that case, the parliament has to be dissolved and a new one elected. The external structure and function of the parliament as a democratic institution remains unchanged even though its internal structure, expressed in its individual membership, may change considerably. The retention of function despite a collapse of structure is a key characteristic of the evolutionary process. In a revolution, on the other hand, the whole system is first destroyed and a new one created without any assurance that the latter will prove functional. In the evolutionary process, however, the breakdown of a given structure is possible only if the new one promises and in fact has higher functional efficiency. The advantage of the new system has to be "demonstrated" before the old one can become unstable.

Apart from the macroscopic changes we have just described, there is the third possibility of indifferent behavior (cf. p. 28). A gradual drift, which corresponds to a purely accidental "regression in time," could be realized only by applying the strategy S_0 described on page 26. We assume that Monod had something like this in mind when he wrote:

> Let me be perfectly clear about this. When I say that living beings as a class are not predictable on the basis of fundamental principles, I do not mean to suggest at all that they cannot be explained by these principles, that they are in some way beyond these principles, and that other principles applicable only to them have to be devised. In my view the biosphere is just as unpredictable as the specific configuration of the atoms making up the pebble in my hand. No one will raise as an objection to a universal theory that this theory does not establish and predict the existence of this specific atomic configuration. We will be satisfied if this real and unique object before us is compatible with the theory. According to the theory, this object does not have to exist, but it may. This satisfies us if we are talking about pebbles but not if we are talking about ourselves. We want to know that we are necessary, that our existence is inevitable and was determined since the beginning of time. All religions, almost all philosophies, and even, to some extent, the natural sciences attest to the indefatigable and heroic efforts of humanity to deny its own accidental origin.[52]

DARWIN: PRINCIPLE OR ISM? An examination of the dynamics underlying processes of selection and evolution shows that the completely unregulated chance situation that Monod celebrates does not occur in evolution. The "fluctuation rectifier" that Monod speaks of represents a phenomenon in which the chance aspect of an elementary event is guided by a natural law that can be formulated in mathematical terms and has macroscopic effects. We are concerned here with the principle—described on page 51 and embodied in the "Selection" bead game—that determines the stability or instability of a state. This principle is just as much a natural law as are the laws of equilibrium thermodynamics. Indeed, it can be derived from those laws, given certain specified constraints or boundary conditions. The fact that instabilities do not generally appear abruptly on the macroscopic level in evolution can be explained purely on the basis of non-stationary constraints that also permit the growth of less well-adapted species.

The fact that the "whole concert of animate nature arose entirely from annoying noises"[52] does not contradict the role that natural law plays in evolution. In equilibriums that bear witness to the laws of thermodynamics, every individual state arises from "annoying noises," too. The difference lies only in the statistical mechanism by which the state takes shape (cf. the Ehrenfest bead game and the game of "Selection"). Seen categorically, but not in terms of its specific constitution, a pebble is just as much the product of natural laws as is a living cell.

In the historical process of evolution, chance has a special role, and we shall have to give more attention to this point than we have so far. Mutation is the source of change, and it always takes rise from a sequence of events that is completely independent of any evaluating mechanism. Monod's example of a pebble illustrates this point. Just as the selection process regulates and controls the effects of mutations, chance—by means of the amplification mechanism provided by natural selection—determines the unique historical and chronological sequence of events. In this context, it is worthwhile to recall Eugene Wigner's statement to the effect that physics is concerned with the regularities of events, not with the more or less coincidental initial or boundary conditions affecting them.

The phenomena of selection and evolution not only can be described as processes that take place under certain environmental conditions and in accordance with certain laws, they can also be reenacted and tested in the laboratory. If experiments of this kind are conducted on representative and relevant systems, the relationship between model and reality becomes immediately obvious. It is crucial, of course, to select representative objects to conduct such tests on and to distinguish principles from accidental factors. We have to let nature show us why certain alternatives were chosen in the course of history. Sidney Brenner once characterized this process by saying: "It's all engineering, molecular engineering."

All that we know in advance when we construct a machine is the function it will have, not its structure or form. The essential task that molecular biology has to accomplish lies in discovering the physical laws that underlie functions. To do this, it first has to experiment with the more or less accidental molecular structures and forms.

Darwin's principle, which seemed at first to attribute to all life processes a special and unique position in the cosmos, appears now to be a principle we can derive from the laws of thermodynamics and state in mathematical terms. But like the laws of thermodynamics, this principle retains its validity only under controlled and clearly defined premises and boundary conditions. This reservation does not exclude the possibility that even where the boundary conditions are not clearly defined, the principle may still provide statements relevant to natural processes.

Darwin himself conceived of his principle as a statement based on experience, and he applied it directly in describing historical reality. It took almost a century to discover the fundamental links between Darwin's central ideas and the quantitative speculations of his contemporaries Clausius, Boltzmann, Maxwell, and Gibbs. François Jacob articulated the correspondence between these ideas in his superb analysis, *Logic of the Living.*[53] Boltzmann's admiration for Darwin knew no bounds. He once remarked, with a modesty approaching self-effacement, that "Someday this century will be known as the century of Darwin."

But Darwin*ism* as such is obsolete. This is not so because his opponents, the vitalists, proved him wrong, for they did not and have not. It is so because a natural law that can be traced back to fundamental principles of physics should not be designated an "ism." It is instead, where its preconditions are met, a law, and it permits of no alternatives. If its preconditions are not met, then it cannot, as a law, lay any claim to validity. Francis Crick[54] has called the anchoring of Darwin's principle in the laws of physics a "foundation of certainty." It provides a foundation for all biological self-organization ranging from evolution to morphogenesis and on up to capacities of the central nervous system for memory.

By tracing the phenomenon of life back to laws of physics and chemistry, we in no way dispute that this new level of organization expresses itself in a form that is typical and characteristic of this level alone, nor indeed do we question that non-material effects can arise from this material order. There is much in this realm that still appears mysterious to us, but the mysteries we have yet to solve can be cleared up through gains in detailed knowledge. Nothing we know about life processes stands in contradiction to known laws of physics, and only the discovery of such contradictions would justify the search for a new and specialized physics of organic processes.

CREATION OR REVELATION? In his lectures, the Göttingen physicist Robert Wichard Pohl often concluded his elucidation of a point by saying: "And that gives us all the more cause for wonder." By this he did not mean that the success of an experiment just conducted was a cause for wonder, although such success might well have appeared miraculous considering what simple means he used to demonstrate the most complex phenomena. The important point is that he made this statement after everything was explained and we might have thought there was nothing left to wonder about.

Wonder is, of course, the source of all inquiry. We begin by standing astonished and helpless in the face of the incomprehensible. Curiosity and a thirst for knowledge grow, the deeper we penetrate into the mysterious darkness and the more facts we illuminate. Once we have gathered these facts, we begin to sort them, compare them, and correlate them until we are able to grasp the overall context in which they belong. But does the solution of a problem necessarily put an end to our wonder? Will the miraculous eventually be researched out of our lives?

In his book *The Pleasure Areas*,[55] the English neurophysiologist Herbert James Campbell describes all the knowledge we have gained in recent years about the localization of pain and pleasure in the central nervous system of higher organisms and about the role they play in directing behavior. Some passages in this book suggest that expanding knowledge is putting an end to wonder. But the exciting results of neurophysiological research in this area by no means indicate that the miraculous will cease to exist if we come to understand it. The understanding of phenomena still does not answer Leibniz's question "Why is there something instead of nothing?"

The central theme of this chapter has been the miraculous order of

living things, order not so much in the sense of order in space and time
—although life does appear in spatial form and in temporal rhythms
—as in the sense of organization, information, and uniqueness. Every
single protein molecule is unique. It was selected from a vastly complex variety of alternative structures and combinations in which the
same building blocks are "systematically" arranged in a different
order and sequence. If we had one example each of all possible protein
structures, we would have so many that they would not fit into the
entire universe even if they were jammed in as tightly as possible. The
fraction of protein structures that have occurred in the entire history
of the earth is so minute that the existence of efficient enzyme molecules borders on the miraculous.

Human beings are quick to categorize the miraculous. They attach
an adjective to it and assign it a place in their world-view:

incomprehensible — God — religion

deterministic — matter — dialectic

accidental — nothing — existentialism

These combinations are by no means fixed. The terms can be
grouped together differently:

God and natural law: "I believe in Spinoza's god, who is manifest
in the harmony of all being, not in a god that is preoccupied with the
fate and actions of men." This was Albert Einstein's answer to a
telegram from the New York rabbi H. S. Goldstein, asking "Do you
believe in God?"

Nothing and dialectic: "We agree that there is no human nature.
In other words, every epoch develops according to dialectic laws, and
men are determined by their times, not by human nature." Jean-Paul
Sartre made this remark in a discussion of his essay *Existentialism Is
a Humanism.* [56]

Jacques Monod rejects—and in our opinion rightly so—any of the
anthropocentric explanations for the phenomenon of life that are
common to most philosophies and religions. In animism, which he
defines as "the consciousness that human beings derive from the

strongly teleological workings of their own central nervous systems and that they project into the natural world," Monod sees a violation of all objective knowledge.

However, the step from an invocation of absolute and blind chance[52] ("Pure chance, nothing but chance, absolute, blind freedom as the foundation of the miraculous structure of evolution . . .") to an a priori rejection of any attempt "to prove, on the basis of thermodynamic calculations, that chance *alone* cannot explain selection in the process of evolution"[57] is not a large one. Such an attempt, if successful, would negate Monod's intention to derive the necessity "for an existential attitude toward life and society . . . from objective scientific knowledge." But is this judgment of Monod's not a new attempt to derive a human-oriented theory of being from the behavior of matter? Would this not be, in other words, a new animism? Otto Friedrich Bollnow characterized this theory in relation to human beings as follows:[58]

> If we try to summarize the anthropological principle of existential philosophy in a few words, we could say that in men there is an ultimate inner core designated, by a term characteristic of this philosophy, as *existence*, which has by nature no permanent manifestation because it only realizes itself in the moment and, in the next moment, ceases to exist in that form. On the existential level, this philosophy claims, there is no permanence in life processes and therefore no preserving beyond the moment what has been achieved, much less any constant progress. There is only an individual leap upward, which grows out of the gathered forces of the moment. It is followed by a plunge downward into the state of drab, uneventful life, which may, at some later moment, give rise to another leap upward.

We should try to free this inquiry into the relative "weight" of chance and necessity from that ideological polarization forced on it by those who expect, from the answer to it, a scientific justification for their point of view.

We have to take two points into account:

1. The relations of order that result from the second law of thermodynamics and its special applications to "open" non-equilibrium systems can only

be statistically interpreted. They refer, in particular, to a partial order (as defined in set theory), one containing alternatives that cannot be compared.

2. The number of possible states is so large that it cannot be "realized" within the spatial and temporal limits of our universe.

These two points make it clear that it is possible to express laws in the form of "relations of order"—such as: "The entropy of a closed system increases as long as the system is not in equilibrium"; or: "A selective value in a limited ecological environment (in an evolution reactor, for example) will achieve an optimum appropriate to the given environmental conditions"—but that the order thus established must be marked by a large number of individual variations in structure. This is so because the total number of possible alternatives is so large that every actual historical sequence in our limited reality has to have its "individual" uniqueness. The law prescribes merely that something will take a certain direction but not in detail how it will do so.

It is precisely such behavior that we simulated with the prototypes of our statistical bead games. But these games cannot even begin to give us an idea of the complexity of reality.

Monod's statement about "absolute, blind chance" applies exclusively to the historically conditioned uniqueness of events. The uniqueness is absolute here because the triggering of a mutation and its evaluation occur on completely different levels, and this fact precludes the possibility of any causal link between the two events. But in spite of this, the fate of a mutant is determined by natural law. The number of "permitted" routes, large as it may be, is relatively small in comparison to the total number of possible routes. The preferred direction imposed by selective evaluation cuts down tremendously the number of possible binary decisions. Natural law here means a channeling, if not a taming, of chance. We human beings too are just as much the product of this law as of historical chance. We are not a product solely of the one or the other.

Someone could perhaps object that this limitation of chance still does not restrict chance enough to make the historical fact of life an event that is probable in principle and that we can reasonably expect to occur within the spatial and temporal limits of the earth or universe. Monod says in this connection: "Our number was drawn in the lottery. It is no wonder that we perceive our existence as special, that we feel like someone who has just won a billion in a lottery."

But we have to object that it was not a matter of hitting the jackpot right off. All that was needed was any winning number. What matters here is the ratio of winning numbers to blanks in the lottery. The competitive behavior inherent in the laws of selection and evolution limited the number of blanks and prevented the majority of them from even getting into the game in the first place. Every winning number meant that the game could proceed for higher stakes. The upshot— the human being—thus represents massive winnings in the game of evolution.

There is considerable disagreement among molecular biologists about how great the probability was for decisive life-creating events to occur. Questions of this kind will be ultimately clarified only by experiment. That does not mean that the whole process of evolution can be reproduced in the laboratory. Such an effort would be pointless from the outset. Scientists design their experiments instead to answer specific selected questions. In the following chapters, we will look at a number of such experiments. With the help of theory, the individual answers gained from experiment can be integrated into a coherent statement, and in this way we can learn which processes are possible and which can be excluded from consideration. The actual chain of historical events cannot, however, be reconstructed in detail.

The most important result of all the experiments conducted on evolution to date (cf. Chapters 13 and 15) has been the discovery of great phenotypical variety on the level of biological macromolecules. This has held true not only for the functional structures, the proteins, but also for the legislative blueprints, the nucleic acids. In every sufficiently large amount of macromolecules synthesized *de novo* there are always variants within the population that are optimally adapted to the environment and whose selection can be repeatedly reenacted in the laboratory. This indicates that matter's capacity for self-organization has been underestimated rather than overestimated.

The probability is greater that "something" rather than "nothing" will happen. This fact lends "necessity" far greater significance than we had previously allowed it and consequently makes the fact of evolution certain or inevitable. This, of course, makes the individual route evolution will take all the more uncertain because this route is now only one of many possible ones.

This is strongly reminiscent of a situation in the statistics of equilibrium states. One of the essential premises under which the laws of equilibrium hold is ergodicity. Ergodicity means that in the course of time every microstate, i.e., every specific constellation of statistical distribution (in whatever approximation), will be reproduced. But if we calculate the periods needed for nearly perfect reproductions of distribution constellations to occur—the physicist calls these periods Poincaré recurrence times—we find that they are usually much longer than the estimated age of the universe, that is, over ten billion years. Boltzmann calculated that for one cubic centimeter of a gas diluted at about one thirtieth of atmospheric pressure, more than $10^{10^{19}}$ years would elapse before the coordinates of all atoms within ten Ångstroms and their velocities within 0.2 percent of their average values would recur.

Another example is the individual differences that all crystals display as a result of flaws. The snow crystals shown in Figure 22 differ from each other because different orientations of the water molecules were possible at the moment of crystallization. A snowflake is made up of over 10^{18} water molecules. If only a billionth of these molecules were out of place in the crystal, there would be more than a billion flaws that could be distributed in $10^{10,000,000,000}$ ways on the 10^{18} lattice positions.

These examples show that any microscopic representation of an equilibrium state has to be regarded as unique in detail. But this individual "historical" uniqueness in no way contradicts macroscopic laws.

The evolution of ideas and the evolution of social systems also follow laws of their own, just as they have their individual and "historical" freedom. If we were to derive an ethic solely from objective knowledge, we should invoke neither the rigid order of a historical world process predetermined by the properties of "dialectic matter" nor the arbitrariness of chance existence.

When we speak of material "self-organization"[59] we are not referring to an a priori dialectical talent of matter, which—if it existed—could not possibly have been known to its early apologists. Marxist philosophers and theoreticians such as Max Raphael[60] go through contortions on this point when they concede that the "spirit" inherent in all matter was initially "not highly developed" but "must already have had a capacity for mapping." And Marxist physicists doggedly invoke an "inner dialectic" of matter derived from its inherent motion.

We want to make it very clear that when we speak of "self-organization" we mean nothing more than the capability of specific forms of matter to develop self-reproductive structures, a capability resulting from definite interactions and combinations that take place under certain constraints. This self-organization is a necessary premise both for evolution and for the development of social systems. But it is in no way sufficient to explain the inevitability of any *specific* historical route. The a priori dogma of a dialectic of matter therefore appears to be a projection of qualities that were acquired at higher levels of organization and by means of superposition and integration under highly specific conditions.

But Sartre too interprets matter "animistically" when he says:[56] "The world of things is merely a source of misfortune. It offers us no handle on itself, it is basically indifferent to us. It is a permanent complex of probabilities. In other words, it is precisely the opposite of what Marxist materialism claims it to be."

"Objective knowledge" will not support a doctrine of nihilism any more than one of social inevitability. Natural science does not claim that men "can do without faith in God" any more than it yields a proof of God's existence. An ethic has to reflect objectivity and knowledge, but it should be based more on the needs of humanity than on the behavior of matter. We also believe that no ethical order can be absolute. It will always have to take different points of view into account and cannot be cut off from its historical roots.

Niels Bohr gave us a clear definition of complementarity: Whether an answer in quantum mechanics will be stated in terms of "waves" or "corpuscles" will depend solely on the line of inquiry in a given experiment. Physics has had to deal with this dichotomy, and it remains a central factor in biology. Life is neither creation nor revelation. It is neither the one nor the other, because it is both at once.

In light of this, we cannot claim absolute validity for an ethic of objective knowledge. Whether we choose to emphasize the "laws of material existence" or the "absoluteness of human existence," a just human order needs for its realization not only objective—and always incomplete—knowledge but also a humanism based on hope, charity, and love.

THREE

The Limits of the Game— The Limits of Humanity

A small ring
Encircles our life,
And races unnumbered
Extend through the ages,
Linked by existence's
Infinite chain

Johann Wolfgang von Goethe,
"Limits of Humanity"

9

The Parable of
the Physicists

The problems we will be discussing in the following chapters will not have been completely thought through until "we have understood what the direst possible turn" our research could take is. In 1974, a group of American biologists decided, for the first time in the history of their science, to call a moratorium on certain experiments.

In the realm of knowledge we have reached the farthest frontiers of perception. We know a few precisely calculated laws; a few basic connections between incomprehensible phenomena, and that is all. The rest is mystery closed to the rational mind. We have reached the end of our journey. But humanity has not yet got as far as that. We have battled onwards, but now no one is following in our footsteps; we have encountered a void. Our knowledge has become a frightening burden. Our researches are perilous, our discoveries are lethal. For us physicists there is nothing left but to surrender to reality. It has not kept up with us. It disintegrates on touching us. *We have to take back our knowledge.*

Möbius, a character in Friedrich Dürrenmatt's comedy *The Physicists*, [61] makes this speech in an attempt to convince his fellow patients Kilton and Eisler, who call themselves Newton and Einstein, to remain in the insane asylum where they are all inmates. The three physicists try to "take back" their knowledge in this way and prevent its misuse. But their effort is in vain: the powers they hope to elude have established a foothold in this very asylum; Dürrenmatt's story takes the "direst possible turn." The physicists' knowledge falls into the hands of a syndicate which then sets out "to rule over countries, conquer continents, and exploit the solar system."

"Every attempt an individual makes to solve for himself problems that affect all of us is bound to fail."

Möbius, Kilton, and Eisler—*nomen est omen*—are among us, and not just in the guise of physicists. For one of them, asceticism is the only possible solution. Another values freedom above all else and is willing to defend it at any cost. A third clings to an ideology. The party that represents this ideology is sufficient guarantee for him that his knowledge will not be misused.

The drama of the physicists is now playing on the world stage. So far it has taken neither the "direst possible" nor the "best possible" turn.

"Every attempt an individual makes to solve for himself problems that affect all of us is bound to fail." Biologists would do well to heed Dürrenmatt's words here. We can easily imagine an insane asylum— perhaps we should call it a neurological clinic—to which a geneticist, a biochemist, and a virologist have withdrawn. They know how to transform cells, transplant genes, cure hereditary diseases. But their knowledge would also allow them to create behavioral monsters. They have therefore decided to withdraw from the world, to work quietly, to keep their knowledge to themselves, and only to help and heal. They think they will be able, in this way, to prevent any misuse of their knowledge.

But can knowledge be kept secret? Would it not be possible for Dürrenmatt's tragicomedy to take place in this very clinic or in any research institute anywhere in the world?

"The content of biology is the concern of biologists, its effect the concern of all men. What concerns all can only be worked out by all."*

*Paraphrased from Dürrenmatt's "Twenty-One Theses on *The Physicists.* "[61]

10

Of Self-Reproducing Automata and Thinking Machines

Molecular biology today has developed highly sophisticated tools for isolating, characterizing, and modifying the "building blocks of life." What we have learned in this process enables us to apply the rules of evolution to "artificial" systems as well. John von Neumann's idea of a self-reproducing automaton might find its first practical application in this area. The analogy to the autocatalytic mechanism of the learning process makes it likely that we will be able to make use of evolution games in developing "intelligent automata." These automata open possibilities to humanity that can contribute to its well-being but that can just as easily be misused.

10.1 "ARTIFICIAL" LIFE?

Will we someday be able to create living beings artificially? If life comes into being according to the laws of physics and chemistry, this is a justified and indeed inevitable question that we cannot dismiss by relegating it to the realm of science fiction.

This problem has two sides to it. We have to distinguish between creating living beings "artificially" and creating "artificial beings," robots that resemble living beings only in certain aspects of their behavior but not in their origin, structure, or shape. We also have to clarify whether we mean living beings in general or, more specifically, human beings. Ultimately, of course, we are thinking of human beings. That is the very aspect of this whole issue that makes it both exciting and frightening.

What is life? In our recent study *Ludus Vitalis*[14] we have examined the definition of this concept in detail and have shown that the best we can do is to list the necessary preconditions that have to be present for a transition from the inanimate to the animate. But a complete description of the concept "life" without reference to a specific living being—or at least to a specific level in the hierarchy of life—is meaningless. The term "life" encompasses a complex variety of phenomena, and it is precisely this variety that we regard as one of the major characteristics of life.

Keeping this in mind, we would like to make two predictions:

1. It will be possible to reproduce *any* living being from its natural hereditary material "artificially," i.e., in a test tube (or breeding apparatus).
2. It will probably never be possible to create living beings *de novo*, beings that will resemble natural beings in every respect, even though differing markedly from natural beings in their hereditary material and structural make-up.

The first statement says that if we understand the mechanisms involved in the formation of cellular structures well enough, we will be able to reproduce and multiply genetic information that has been isolated from the cells of certain individuals—possibly introducing planned changes into that information—and to transcribe this information back into real "living" structures.

There is no reason why we should not be able to initiate non-sexual reproduction, e.g., by transplanting[62] diploid nuclei of somatic cells from a male or female donor into denucleated egg cells. In this way, we could produce any desired number of genetically identical individuals. The creatures produced "artificially" in this way from individual molecules would, of course, be almost indistinguishable from their fellows.

The thought of applying this process to human beings is a frightening one—the stuff of horror films. In his essay "Molecular Biology and Metaphysics," Gunter Stent sums this up in a graphic way: "It might be fun to have Marilyn Monroe as a neighbor, but the prospect of meeting her a thousand times over wherever one went in town would be enough to give anyone nightmares."[63]

Such speculations still belong to the realm of science fiction, but there are questions of gene transplantation and manipulation that we do have to deal with now.

10.2   GENETIC MANIPULATION

The genetic information that makes up the "building plan" of an organism is contained in the chromosomes in the form of molecular "printer's plates." For a human being, this set of plates could print out the equivalent of a sizable private library. The central problem in gene transplantation is locating and exchanging specific pieces of information in this vast amount of material. The gene map—the pattern in which individual genes are arranged—as well as the syntax and semantics of this molecular language are, as we will show in Chapter 15, fairly well understood today, at least for primitive organisms. The difficulties of gene transplantation are primarily technical ones, problems of "engineering." Gene transplantation is comparable to organ transplantation, and the problems encountered in both are similar. The major difference is that with gene transplantation we are working in molecular dimensions and therefore need microinstruments capable of functioning in this realm.

At first, molecular biologists seemed faced with a hopeless task, but nature came to their aid. Men did not have to invent gene transplantation; all they had to do was discover it in nature. The required molecular apparatus already existed. Scientists "merely" had to isolate it from the natural sources. A genetic operation involves three phases:

1. The precise excision of either a partial or a total sequence of a gene from a donor chromosome and, usually, from a receiver chromosome as well.
2. The "packing" of the transplant and the introduction of it into the receiver organism.
3. The implantation of the transplant into—i.e., the fusion or stabilization of symbiosis with—the receiver chromosome.

All these problems have, in the meantime, been solved in principle. A number of years ago, Werner Arber at the University of Geneva and Matthew Meselson and his colleagues at Harvard University discovered an enzyme of particular importance for operations 1 and 3. Today, we are familiar with a whole class of such "restriction enzymes" that can be used to subdivide genetic material in precise ways. But the practical application of these enzymes could be explored only after various scientists in the United States—primarily Herbert W. Boyer, Stanley N. Cohen, and Matthew Meselson—had investigated the mechanisms by which they work.

A restriction enzyme makes its "cut" in the gene's double chains of nucleic acid at a specific site characteristic for it. The site consists of a palindrome of usually six letters in the genetic alphabet (see Figure 37). A few years ago Walter Gilbert found that symmetrical sequences were also used as start signals for gene-copying. Although we still do not fully understand the recognition mechanism itself, it would seem that palindromic sequences are the general means used to mark genetic information for recognition by the executive component in proteins.

The way in which the double chains of nucleic acid are cut (see Figure 37) is crucial for the functioning of the restriction enzyme. The fragments that result from the cutting process have dangling chain

Palindrome:

HAN|NAH

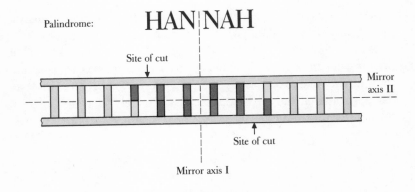

Site of cut

Mirror axis II

Site of cut

Mirror axis I

GENE FUSION

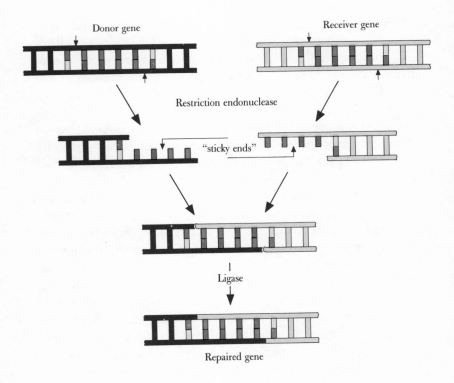

Donor gene

Receiver gene

Restriction endonuclease

"sticky ends"

Ligase

Repaired gene

ends, so-called sticky or cohesive ends. They are called sticky or cohesive because they tend to reunite with corresponding complementary ends wherever they encounter them and are fused together with them by appropriate enzymes, which are known as ligases.

The principle is clear. The restriction enzyme cuts out the genetic material in both the donor and the receiver at the same specifically marked site. If the excised piece is transferred from the donor to the receiver, the sticky ends of the transplant will automatically be fused into place. Since the molecular language of genes makes use of four different "letters," a large number of different cohesive recognition signals are possible, depending on the length of the symmetrical recognition sequence. In restriction enzymes, the molecular biologist has at his disposal a versatile instrument for the exact division of chromosomes.

The problem of transfer has also been solved. Biologists have known about the so-called plasmids for some time; Joshua Lederberg discovered them in 1953. Plasmids are small packets of genetic material that lead an autonomous existence in bacterial and other cells but that can sometimes also unite with the cell nucleus and integrate their information into the genome. These particles are of great importance in cell transformation and fusion. Transformation means that a cell is irreversibly changed. A familiar example is the transformation of a normal cell into a malignant cancer cell. Fusion means that genetic

Figure 37. The function of a restriction enzyme in gene transplantation. The upper part of the illustration shows two palindromes, one in our language and the other in the language of nucleic acids. The lower part shows the individual reaction steps in gene transplantation. The restriction enzyme recognizes a palindromic symmetry in the genetic "print." It then cuts both chains off at precisely determined intervals and leaves two cohesive chain ends extending. Because this process occurs in an unequivocal way, all DNA chains cut by an enzyme of this kind are alike; hence donors and receivers can be fused, yielding an intact genetic message. The enzyme that fuses the complementary ends is called a ligase.

material from two different donor cells, cells from a mouse and a rat, for instance, is united in a viable hybrid.

In gene transplantation, the plasmids are utilized in two steps. With the help of restriction enzymes, the donor gene is integrated into the plasmids. Then the plasmids are used to introduce the gene into the receiver cell (see Figure 38). In nature plasmids have a similar function. They are, for example, responsible for gradually making bacteria resistant to antibiotics, such as penicillin. The bacterial genome itself lacks this flexibility and can attain it only through a symbiosis with plasmids that have the necessary capabilities and that may transmit them to the bacterial genome. In this way, resistant strains can reproduce themselves selectively.

The possibilities that these insights open up are vast. There is little doubt that the mystery of cancer will be at least partially solved by research in this area. We may also be able to cure congenital diseases, such as the mental deficiency associated with phenylketonuria; and we may be able to discover new sources of food, perhaps by a "cell-free" production of protein. It is quite conceivable that by combining genetic materials from different sources experiments in evolution may produce new kinds of beings. We are not thinking so much here of science fiction monsters as of "protein factories" or other useful organisms. Our imagination is probably too limited even to begin to picture what the future may bring.

But as soon as we make practical use of our knowledge we no longer stand beyond good and evil. We have to realize that our actions could take either the "direst possible" or the "best possible" turn in Dürrenmatt's sense.

Figure 38. The transportation mechanism in gene transplantation.[64] Plasmids live in symbiosis with cells. If a cell division takes place, they are reproduced along with the genetic material (cell nucleus or its equivalent). We say that a cell that has undergone some new symbiosis is transformed. (Plasmid information is often fully integrated into the nucleus.) Figure 37 shows how plasmids use a restriction enzyme when they act as vehicles for the transplantation of genes.

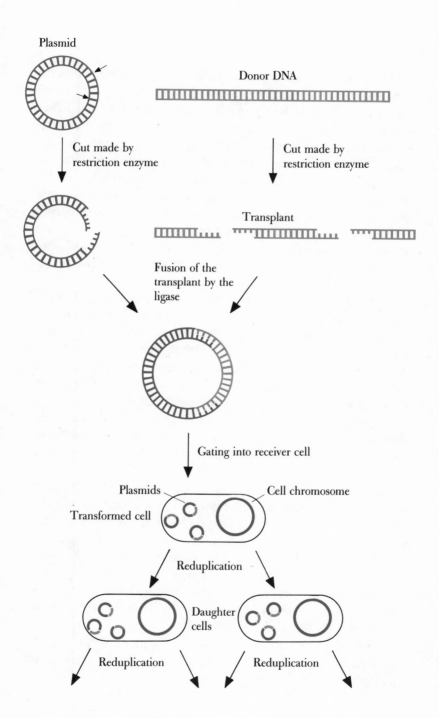

Plasmid

Donor DNA

Cut made by
restriction enzyme

Cut made by
restriction enzyme

Transplant

Fusion of the
transplant by the
ligase

Gating into receiver cell

Plasmids

Cell chromosome

Transformed cell

Reduplication

Daughter
cells

Reduplication

Reduplication

No one would refuse to heal if healing were possible. On the other hand, we have to minimize all risk and prevent any conceivable misuse of knowledge. This is a problem that "concerns us all." There is always a risk when we cannot foresee the consequences of an experiment. It is not enough just to reject all "experiments" with human beings. We must also see to it that any experiments we conduct do not, even indirectly, have harmful effects on human beings or their environment. Pathogens exist in our natural environment, but as a rule they are under control, precisely because men have experimented with them. We cannot take the risk of producing new pathogens whose effects would be totally unknown and against which we could be utterly defenseless. At the same time, however, we can prevent misuse only by knowledge.

It is remarkable that a number of American biologists, led by Paul Berg, initiated public debate of these issues[65] and called a moratorium on research in gene transplantation until these public issues could be clarified. Along this same line, the British Parliament appointed a committee to define possible dangers and establish policies for avoiding them.[66] These are both attempts to act reasonably and responsibly outside the arena of obvious political polemic.

Simply making problems public, however, does not guarantee that they will be solved in the light of the best knowledge and the highest moral standards. Many problems require detailed knowledge that no plebiscite can replace.

In its efforts to conquer sickness, hunger, and poverty, humanity will be grateful for the knowledge gained from molecular biology. But we must not lose sight of the dangers involved. Whoever wants to prevent the misuse of knowledge cannot ignore knowledge and leave it in the hands of those who want to misuse it. Moratoria on research can be useful for clarifying certain issues, and they may become necessary to avoid known risks. Absolute prohibitions, however, apply only to the application or misuse of knowledge, not to knowledge itself.

Man has been driven out of paradise and is forced to "eat from the tree of knowledge." New knowledge demands new ethical and moral standards.

10.3 INTELLIGENT AUTOMATA

Does not our claim that there never will be artificially created organisms comparable in perfection to human beings go against the history of scientific discoveries and inventions? Have not claims of this kind been proved wrong time and again almost as soon as they have been made?

We should use the word "never" with caution. Perhaps we should "never" use it. The same is true of the word "everything." And we should not conclude from our experience that we will ever know "everything," much less be able to do everything. On the contrary, as our knowledge increases, the limits of what we can do will become clearer.

On the wall of a lecture room in the physics department in Göttingen, the following sentence is prominently displayed:

Simplex Sigillum Veri !

But it is only abstractions from reality that are simple. Principles are simple, but reality presents itself to us in its full complexity. The more of reality we comprehend, the more aware we become of horizons that keep receding before us and are as unattainable as the peripheries of the expanding universe.

Many problems have long been solved "in principle." The principles of thermodynamics have been thoroughly understood for about a hundred years, but we still cannot predict the weather very accurately. It is always the details that trip us up.

Our generation will learn more and more about the complicated functionings of our brains, but we will not be able to reconstruct a brain in detail even on paper.

We have no adequate idea of the complexity of real living structures. It lies beyond our comprehension. Even astronomical dimensions fail to convey it. The code text of the hereditary building plan of a single coli bacteria cell consists of about four million symbols. This means that to create such a cell a choice has to be made among about $10^{2,400,000}$ (a one followed by 2,400,000 zeros) alternatives.

This code text is no more produced by "chance" than a literary masterpiece is achieved by tossing letters together.

The human brain is a network of more than ten billion nerve cells, each of which has several thousand points of contact with neighboring cells. Behind this complexity there is a principle of hierarchical organization that scientists the world over are working feverishly to understand.

In 1936, the English mathematician Alan M. Turing asked whether it would be possible to construct a universal thinking machine that would not only calculate the values of given functions according to precise instructions but would independently work out mathematical procedures, so-called algorithms. By storing these algorithms and making them the starting points of new operations, this machine would be able to arrive at any possible "computable" function in a finite number of steps.

The crucial element in such a machine is the storage system—the so-called Turing table. All calculation instructions are entered on it, both those programmed from the beginning as well as the algorithms worked out by the machine. This Turing table thus represents a practically unlimited computing tape. The machine itself carries out only four operations. It moves to either the right or the left on the tape, prints on an empty space on the tape, and stops. All the intelligence lies in the instructions printed on the tape. Similarly, in the normal computers we use to calculate the values of given functions, all the intelligence is located in the program devised by a human brain, in what we call the software.

Creativity in mathematics does not lie in the mere application of known mathematical principles but in the intuitive discovery of new algorithms. As soon as such algorithms are recognized as generally valid and have been incorporated into a logical scheme, their use is reduced to a predictable, routine process that can be best carried out by a machine. The human mind, which is motivated by playful curiosity and takes pleasure in imaginative combinations, is quickly bored by monotonous activities.

It is not surprising that the Turing machine still remains an abstract idea. Its realization would require us to think it through consistently ourselves first. Mathematicians are fascinated primarily by the fundamental problems involved in creating such a machine: the computability of functions in a finite number of steps, the countability of the values of functions in a given domain, and the decidability of problems (i.e., whether we can decide, after using an algorithm for a finite number of steps, if it will yield the desired result). The fact that there are problems in mathematical logic that cannot be decided imposes fundamental limits on the Turing machine, limits that may reflect only the finite and imperfect nature of the human mind. Perhaps the most significant step toward realizing the Turing machine was John von Neumann's idea of a "self-reproducing automaton." This was primarily an abstract model that could be precisely formulated in mathematical terms, but it was based on a number of presuppositions that have also been recognized as essential to the self-organization of living organisms.

Every machine makes use of free energy. It runs on electricity or is powered by an internal-combustion engine. In short, no machine can work without this metabolism. One task the von Neumann automaton would perform is self-reproduction. The first model, designed in 1950, was conceived of in thoroughly realistic terms. The machine would move around in a large warehouse and gather together all the parts necessary for its reconstruction. The most important point is that it would also reproduce its own blueprint because its descendants would also have to have the capability of reproducing themselves. This self-reproduction of the blueprint would make it possible for the von Neumann automaton to perfect itself, and theoreticians were quick to explore this possibility. Selective changes in the program could produce constant improvement in the machine and an extension of its capabilities. This process is comparable to Darwinian evolution.

Making use of work done by his colleague Stanislav Ulam, von Neumann was able to refine his calculations and make them more generally applicable. Von Neumann's mental experiment, which we can easily present in the form of a game, makes use of a homogeneous space subdivided into cells. We can think of these cells as squares on a playing board. A finite number of states—e.g., empty, occupied, or occupied by a specific color—is assigned to each square. At the same

Table 10. J. H. CONWAY'S BEAD GAME ''LIFE''

The playing board for this game should have as many squares as possible. Coordinates are not necessary. Beads of two different colors should be available in sufficient quantities. (A frame that can be centered around any given square can be of help in pinpointing neighborhoods.) Both orthogonally and diagonally adjacent squares make up a neighborhood in this game.

The game is played in distinct phases that we could call generations. Within every generation the rules are applied simultaneously to all squares. The rules determine whether a square will be occupied or cleared. There are only two alternatives. A square has to be either empty or occupied by a bead.

The rules are:

1. Survival: A bead survives to the next generation if two or three neighboring squares are also occupied. The bead then remains in its position.
2. Death: A bead is removed if more than three or less than two neighboring squares are occupied. In the first case, the system is overpopulated; in the second, the individual is too isolated.
3. Birth: A bead can be placed on an empty square if three and only three neighboring squares are already occupied.

In each generation, every square in a configuration undergoes the change the three rules dictate for its state. It is therefore useful for the sake of clarity to use beads of two different colors. The board is inspected for empty squares, and these squares are filled with beads of the second color (white) if the conditions of Rule 3 are met. Then—and this step is still part of the first generation—the squares occupied by black beads are located and transformed according to Rules 1 or 2. The squares occupied by white beads are, of course, still considered "empty" in this step. When all the changes within one generation have been carried out, the white beads are then replaced by black ones. The life of one generation is thus concluded, and we can go on to the next generation, in which the steps listed above are repeated.

The game begins when the player (or players) has placed an arbitrary number of black beads—say, six to ten—in a configuration. This is the only opportunity a player has to influence the course of the game. All further steps are predetermined by the rules.

If the game is to be played competitively, two players can divide the board in half, and each player places his beads on his half. Both initial figures have to contain the same number of beads. In scoring, the number of

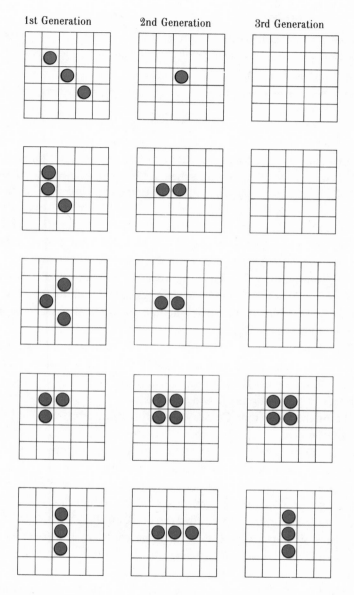

Figure 39. The fates of some triplet configurations in Conway's "Life" game. The first three triplets die out after the second generation. The fourth triplet forms a stable block, and the fifth—called a blinker— oscillates indefinitely.

generations that each player's figure survives and the number of beads placed on the board in each generation are counted. These two sums are then multiplied together.

---

time, a neighborhood is defined for each cell. This neighborhood can consist of either the four orthogonally bordering cells or the eight orthogonally and diagonally bordering cells. In the space divided up in this way, transition rules are then applied simultaneously to each cell. The transition any particular cell undergoes will depend on its state and on the states of its neighbors. Von Neumann was able to prove that a configuration of about 200,000 cells, each with 29 different possible states and each placed in a neighborhood of 4 orthogonally adjacent squares, could meet all the requirements of a self-reproducing automaton. The large number of elements was necessary because von Neumann's model was also designed to simulate a Turing machine. Von Neumann's machine can, theoretically, perform any mathematical operation.

The abstract idea of a cell automaton can be illustrated by a game that the English mathematician John Horton Conway invented and called "Life." With this game, we can simulate rises, declines, and changes in populations of living organisms.

We might think that Conway chose his rules arbitrarily, but that is not the case at all. The game was devised not only to present realistically the effects of isolation, cooperation, overpopulation, etc., but also to be as interesting a game as possible. If the game is to be interesting, however, fates of populations should not be obvious from the outset. For this reason, there should be configurations that can grow indefinitely and others that die out quickly.

Figure 39 shows typical fates of some triplet figures (trominoes). The temporal sequence of a number of generations appears here as a spatial sequence—a mode of presentation we shall make use of often. The first three figures are unstable and die out after two generations. The fourth figure forms a stable block, and the fifth one oscillates at intervals of two generations.

Figure 40 shows a pattern that emerges from a tetromino after ten generations. There are also periodically changing structures that repeat themselves after a number of generations but that move a certain number of squares in the process. Conway's "Glider" (Figure 41) is a typical example. It is an oscillator that reverts to its original shape after a period of five generations and moves diagonally one square to the right and one square down on the board. The maximum speed at which a configuration can move in this game is the same as that of a king in chess. We could designate this speed the "speed of light" because it represents an absolute upper limit. We could then say that the "glider" moves across the board at one-fourth the speed of light.

In an extensive article in *Scientific American,* [67] Martin Gardner described a number of configurations, many of which are characterized by scurrilous behavior. The reader may want to try out some of these figures.

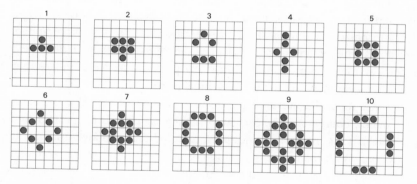

Figure 40. The development of a tetromino.

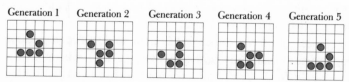

Figure 41. The "glider" is a figure that repeats itself after four generations but, in the process, moves one square down and one square to the right.

Two types of configurations make "Life" into a strategic game. Figures of the first type are called "guns" because they shoot off projectiles; those of the second type are called "eaters" because they do just what their name indicates: they swallow up other figures without undergoing any changes themselves. An interesting story is connected with the discovery of these two kinds of figures. Conway had speculated that as a result of overpopulation regulation there were no figures that could grow indefinitely. But since he could not substantiate this hypothesis, he offered a reward of fifty dollars to anyone who could prove or disprove it. (In mathematics a hypothesis is considered disproved if only one example can be found that contradicts it.) The prize went to a team of young scientists working on the "Project of Artificial Intelligence" at M.I.T. They simply set the computer the task of finding an example that would disprove Conway's hypothesis. The computer came up with the so-called "glider gun" (Figure 42), which fires off a "glider" at regular intervals. The "gun" itself is a stationary or spatially fixed oscillator that resumes its original shape after thirty generations. In each cycle, it emits a "glider" that wanders across the playing board, and since the "gun" oscillates indefinitely, it can produce an infinite number of "gliders." And that means that there *are* figures that can grow indefinitely.

All these discoveries have inspired a state of veritable euphoria among computer engineers and mathematicians. Both professional journals and newsmagazines have reported these discoveries in great detail. With the aid of the computer it is possible to stage large and strategically sophisticated contests and to discover new and increasingly bizarre combinations.

But the idea behind this game is a serious one involving the development of a new experimental mathematics. Mathematicians hope that these experiments will help them someday develop a new generation of computers that will realize Turing's idea.

In one respect, of course, the name "Life" is not justified. The game does demonstrate beautifully how simple rules can result in an extremely complex "real world," complex not only in terms of its spatial structure, or morphology, but also in terms of its functional behavior. The game does display many qualities that are reminiscent of life in

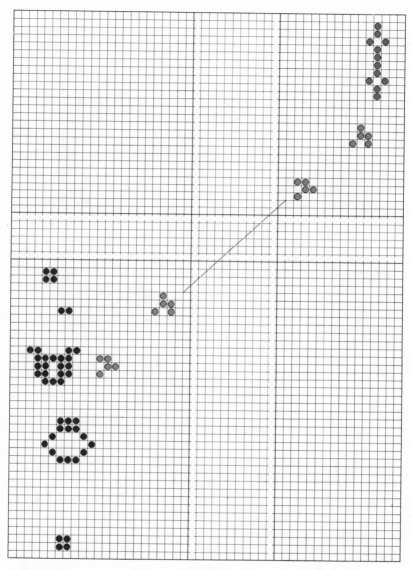

Figure 42. The "gun" configuration (shown in black) is a spatially fixed oscillator that resumes its original shape after thirty generations and, within this period, emits a "glider" (red). The glider wanders across the playing board and encounters an "eater" (green), which happens to be a fifteen-generation oscillator. The eater swallows up the glider without undergoing any irreversible changes.

both a positive and a negative sense. But one crucial characteristic is lacking, and that is "creative" chance. The course that Conway's game of "Life" takes is a classic example of a deterministic process. Everything evolves with strict causality from the initial conditions, and behind it all there is a "fully informed creator." The game can do nothing but enact his ideas. No allowance is made for chance.

This game cannot, of course, convey the immense complexity of living organisms in nature in terms of the actual physical processes involved. If the rules of evolution turned out to be identical with Conway's rules, this would prove the existence of a fully informed creator and would thus constitute a proof of God's existence. But all the knowledge we have assembled to date on the molecular structure of organisms and on their evolutionary changes speaks against this simplistic conclusion. It would make no sense to propose a God who stands in conflict with the laws of physics. Such a God would stand in conflict with Himself. In a series of recent interviews, Eike Christian Hirsch[68] summarized the contributions that physics and biology since Kant have made toward a "proof of God's existence." Hirsch concludes:

> We no longer believe that the origin of life is some mysterious place where God is more present than elsewhere. Either we believe in God everywhere or nowhere. The origin of life offers no proof of God. [See also Georg Picht, "Der Gott der Philosophen" ("The God of the Philosophers").]

Conway's game represents nothing but deterministic average behavior in the dynamics of a genetically invariable ensemble made up of a large number of individuals. We can, of course, introduce chance into this game, but then the results will be identical with those of the selection games described in Chapter 5.

In his book *Evolutionsstrategie,*[70] Ingo Rechenberg has shown that a trial and error method employed in a multidimensional space representing all alternatives in decision-making will yield results much more rapidly than a deterministic search based on checking the gradients with respect to all the coordinates of the space. Having seen in Chapter 8 how molecules can organize themselves and make selective use of functional properties, we can understand here how a purposeful development can be set in motion in a complex, multidimensional phase space even though that development is not guided by a plan worked out in advance.

If we try to imagine today how a Turing automaton could be constructed, we would assume that it would have to be conceived along the lines of a living organism. This statement would seem to contradict, at least partially, the second prediction we made at the beginning of this chapter. The crucial point is, of course, that we would not be able to construct this automaton. It would have to construct and organize itself, and it seems certain that the interplay of chance and selection in this case—if for no other reason than the complex variety of alternatives available—would produce something very different from any known living organism.

What would such a machine look like? First of all, it would have to have the features that von Neumann described in his original idea of a self-reproducing automaton. In other words, it would have to have a storage system large enough to enable it to develop algorithms. Then it would also need adaptive capabilities that would allow it to reorganize and expand its program as it functioned. But as we showed in Chapter 5, that is possible only if the machine has a built-in means of evaluation.

In the evolutionary process, only later generations benefit from improvements in blueprints. What exists is not modified itself; it simply dies out. In terms of machinery, this would be a prohibitively expensive process. Nature's use of receptors that take in signals from the environment and her development of nervous systems that interpret and store those signals point out a more economical route. Learning, too, is a selective process involving the reproduction, evaluation, and modification of elementary processes that take place in the learning entity. It is clear that new ideas arise at the expense of old ones. But all that needs to be discarded is the old ideas, not the entire machinery from which those ideas originated.

The Turing automaton will therefore have to have an inherent, automatically functioning evaluation mechanism that will motivate it to do some things and not to do others. It will have to have pain and pleasure centers, will have to be able to experience fear and joy. That, at any rate, is how an "animate" being learns.

There is no doubt—now that we are certain of the existence of such centers in the brains of higher organisms and have localized some of them—that we will soon understand the principles by which they function.

The fact that we can explain a phenomenon in terms of physics does not mean that that phenomenon ceases to exist or that any of its miraculousness is taken from it. Research in these areas will in no way strip man of his individual qualities and degrade him to an entity understandable in terms of calculable quantities. Indeed, for the thinking man there is great comfort in the realization that his existence is embedded in the unity of nature, that his life is an episode in the universal history of life. He is not a lonely "gypsy wandering on the fringes of the universe"[52] but is instead part of a totality; and, indeed, each individual is, because of his consciousness, the center of that totality.

Is it reasonable to devise machines that will have what will probably be only a very limited capability for self-reflection? Is it not much more important to focus our efforts on human society, which is the next evolutionary step above the individual, and to shape that society into a rational organism that will stop destroying itself?

11

"From One
Make Ten . . ."

*The human population explosion represents hubris in the game of life.
In nature we find different forms of growth that we can classify on the
basis of normed strategies. Exponential growth results from an autocatalytic reproductive mechanism. The doubling time of a species is
a constant that is characteristic for that species. If we make use of birth
control, we can affect the amount of time it takes for the human
population to double. At present, this doubling period is shrinking
constantly, and the world's population is burgeoning in accordance with
a hyperbolic law. The potential for catastrophe is great. The self-
regulating mechanism that will, because of the inherent limits of our
environment, inevitably go into effect is inhuman. This mechanism
would result in a devastating death rate.*

11.1 RATE FUNCTIONS AND LAWS OF GROWTH

In the first sixteen centuries of our era the world population increased from about two or three hundred million people to about five hundred million. In the next two hundred years it rose to a billion. In the course of the next hundred years, it doubled again, reaching the two billion mark sometime around 1930. Today, there are nearly four billion people living on the earth (see Figure 43).

If in successive time intervals of the same length a number doubles itself or multiplies by the same factor, we have an example of an exponential law of growth. Self-reproduction or autocatalysis underlies this law. The quantity at hand—whether we are dealing with neutrons in a block of uranium, bacteria in a culture, people, capital, information, or knowledge—will catalyze, program, and regulate its own reproduction.

The population of the earth at the moment is experiencing greater than exponential growth. The birth rate is far outstripping the death rate, and at the same time the average number of births per inhabitant, particularly in the less developed countries, is still increasing. The reduction of infant and child mortality has raised average life expectancy and, consequently, increased birth rates. The time it takes for the world's population to double is constantly becoming shorter. At present, it is shorter than the lifetime of a human being, and it is reasonably accurate to say that each generation brings a doubling in the world's population.

The mathematician describes growth either by specifying the cumulative changes or by specifying the changes per unit of time—the rates. Here there are two fundamentally different methods of representation. We can deal either (1) with the change in a quantity as a function of time—we call this the *law of growth*—or (2) with the dependence of the rate of growth on the amount of the growing quantity—we call this the *rate function.*

If the rate function is constant—i.e., the rate or change per unit of time does not depend on the quantity—then the total quantity will

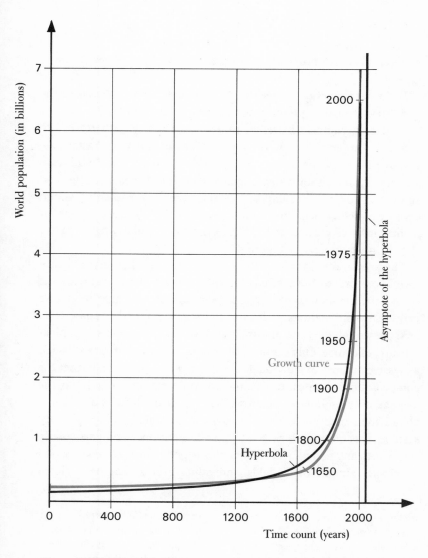

Figure 43. Growth curve of the world's population,[75] extrapolated to the year 2000 (red curve). The hyperbola drawn in black corresponds almost exactly to the population curve over a long period of time. The hyperbola disappears into infinity in the year 2040 (see the asymptote). The graph vividly illustrates, on the one hand, the threatening nature of the population explosion; but it also shows, on the other, how meaningless extrapolations can be, particularly if they are made in the neighborhood of singularities.

increase in direct proportion to time, and we have a *linear law of growth.*

If the rate function is linear—i.e., the rate is proportional to the quantity—then the system grows *exponentially* with time.

If, finally, the rate increases more than linearly with respect to quantity—in proportion to the square of the quantity, for instance—we have a "superexponential" or *hyperbolic law of growth.* In this case the time spans within which a quantity doubles become shorter and shorter. If no limits are imposed on this process, the quantity would become infinitely large after a finite period of time. These three fundamental laws of growth and the rate functions underlying them are shown in the graphs of Figure 44.

The precondition for growth of any kind is that the births or new formations exceed the deaths or decompositions. If both rates are the same—if formation is exactly compensated by decomposition—we call the system stationary. Its size undergoes no further changes.

What we call the rates of formation and decomposition in a population are, of course, only statistical averages reflecting the rate at which a massive number of individual events take place. This means that the compensation of rates in a stationary state can occur only on the average over time. Here again we have to deal with the question—already discussed in Chapters 3 and 4—of the stability of a stationary system. This question is of great general importance and particularly relevant to any policy-making that affects population, currencies, and economics. In all these fields, individual events that have global

Figure 44. Here three different laws of growth (which show the dependence of the quantity on time) are compared with their underlying rate functions (which show how a quantity's change per time unit depends on the quantity itself). Growth curves are characterized as follows: *Linear* (top), quantity increases by the same amount in equal time units. *Exponential* (middle), quantity increases by the same factor in equal time units. (Despite this snowball effect, the quantity will reach infinite proportions only after an infinite amount of time.) *Hyperbolic* (bottom), the doubling periods shrink with time. (The quantity becomes infinite within a finite period.)

RATE FUNCTIONS   LAWS OF GROWTH

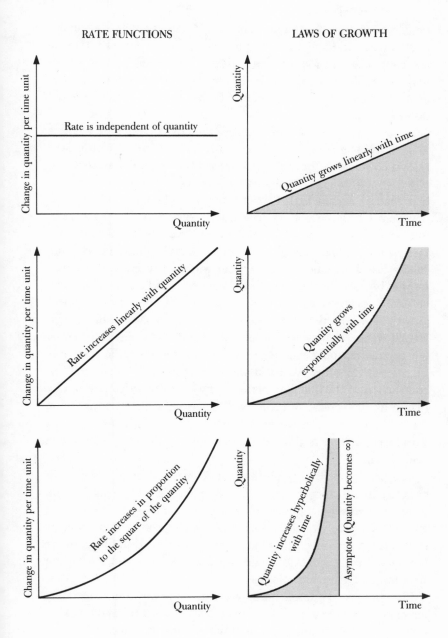

ramifications are autocatalytic in nature. Measures designed to inhibit or stimulate can plunge a system into considerable danger if they do not contain self-regulating mechanisms that respond to the effects they evoke. The key question here is how the size of a population or its fluctuations affects the difference between rates of formation and disintegration.

Let us assume that this difference is directly correlated to changes in quantity. The simplest case occurs when the difference between the rates is constant on the average. If the rate of formation exceeds that of decomposition, what is added will always be the same amount in time spans of equal length. (This relation should not be confused with the one in which the amount present is doubled or multiplied in equal time spans.) Linear growth of this kind can be controlled with relative ease as long as the external conditions bearing on it are understood. In this case, fluctuations that do occur will not trigger internal catastrophes because these fluctuations do not snowball and therefore can be easily compensated for. But we cannot conclude from a linear growth alone exactly what kind of mechanism is causing it, particularly if we are dealing with a limited span of time. The exponential law of growth, too, has a nearly linear phase in its initial stage (see Figure 44). In other words, if the time span observed is small in comparison to the doubling period, we will always register no more than a linear increase in quantity. What is immediately threatening for humanity at the moment is that the time span it takes for the world's population to double is about equal to the time span of one generation.

Internal compensating mechanisms can cut exponential growth back to a linear progression. The increase of self-reproducing bacteria in an inexhaustible medium is truly exponential, but the vertical growth of a tree, which is also the result of cell reproduction, is controlled in such a way that the overall growth is about linear except during the early stages of rapid cell accumulation and during the slackening of growth in old age.

In many social games, linear growth conditions are deliberately maintained over long periods of play. Only toward the end do exponential changes go into effect. If this were not so, it would be clear at the beginning of the game who the loser was. He would then have to play this role to the bitter end, and that could be frustrating. The world-famous Japanese game of go and go-bang, the version of it developed in England, offer classical examples of how catastrophic situations can gradually evolve out of a linear course of play.

Go demonstrates once again how complexity can arise from the repeated application of the simplest principles. The rules of the game are so simple that anyone can learn them easily and put them into practice immediately. But even the most gifted player can never achieve complete control. In Japan, a special branch of science has developed around go, and until 1868 there was even a special go academy.

A match between equally skilled players can go on for days. Because of the linear increase of beads on the board, it can remain unclear for a long time which player will win. The strategy of enclosure introduces a nonlinear element into the game and eventually leads to a dramatic turn in events. The deviation from linear growth is quite complicated in its mathematical formulation. The rules were not, after all, conceived of mathematically. They evolved historically and are based on a strategical concept. The mathematical structure of the games discussed in Chapter 5 is, by comparison, much simpler. Play in these games is dominated almost exclusively by laws of growth that are purely exponential, and we will now investigate these laws more closely.

A purely exponential law of growth goes into effect whenever the difference between the rates of formation and decomposition is positive and directly proportional to the basic quantity involved. If this quantity is made up of individual units, their number will keep doubling at regular intervals. This doubling period is a constant that is characteristic for the system in question. The individual doubling process may be discontinuous, but if we consider a large enough number of individuals, the doubling process becomes a practically continuous one that can be adequately described by a continuous function. This function has the property, dictated by the rate function, that the derivative with respect to time will show the same dependency on time as the function itself does. Instead of citing the doubling period, we often give the time it takes for a quantity to increase e

Table 11. BEAD GAMES "GO" AND "GO-BANG"

Go—or I-go, as it is called in Japanese—is a board game for two players. Each player has 181 pieces, either black or white. The board is divided by 19 vertical and 19 horizontal lines that form 361 intersections on which the players place their pieces.

Go-bang can be played by two or more players on a board of 13 by 13 squares. All versions of the game begin with the players taking turns placing beads of their color on any empty squares they choose.

In go-bang each player tries to place his beads in chains of five. The chains can be vertical, horizontal, or diagonal, but there cannot be any gaps in a chain. The player who first forms a chain of five beads is the winner. The game can also be played until the whole board is filled. In this case, the player who has the most chains of five is the winner. In a third version, whenever a player places the fourth bead in a chain, he may remove an opponent's bead from the board. Whoever removes the most opposing beads is the winner.

Mini-go is played exactly like go but on a smaller board of 13 by 13 lines. The rules are as follows:

1. The players take turns placing their beads on the board. These beads remain on the board until they are enclosed and "captured" by the opposing player.
2. Beads in fully enclosed regions can be "captured" and removed from the board, but the players are not obliged to utilize every possibility to enclose the opponent's beads.

The application of the second rule presents no difficulties as long as a player is dealing with completely occupied and enclosed regions. In situations on the edge of the board, the opponent's beads do not have to be fully encircled but can be merely trapped against the edge. If the enclosed regions are not completely filled by an opponent's beads or if there are possibilities of forming overlapping enclosures, then a player must decide what to do on the basis of the types of configurations shown in Figure 45. The game is over when all winning positions have been occupied. In scoring, each player counts up the number of empty positions surrounded by his beads and subtracts from it the number of his beads his opponent has captured. Whoever has the highest number of points wins.

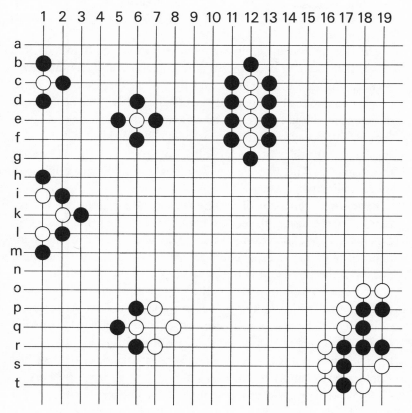

Figure 45. The most important situations and concepts in go. In the two situations in the upper left a white piece is completely enclosed by black ones and is therefore captured by black. Several pieces can be captured at once if they are in a chain, as shown in the upper right. In the configuration shown on the left-hand side of the board in the middle, it would be disastrous for white to occupy the free position k1. But by occupying this position, black can capture all three white pieces. In the figure in the lower center, black can capture the white piece on q6 by placing one of his pieces on q7. But by reoccupying q6, white can capture the black piece on q7. This seesaw process could go on indefinitely. For this reason the rules stipulate that a piece that has "killed" a single other piece cannot be removed as an individual by the next move. In the lower right-hand corner the "seki" situation is shown. None of the three empty intersections can be occupied by either black or white.

times. The transcendental number $e = 2.718$ is the base of the exponential function.

The principles controlling a growth curve also apply to a decline curve. If the rate of decomposition dominates, and if the difference between the rate of formation and that of decomposition is again proportional to the basic quantity involved, then the time dependence of the quantity will be given by an exponentially decreasing function. Radioactive decay is a familiar example. The so-called half-life of a substance is the time it takes for half the atoms of the original amount to decompose. This half-life can be thought of as a typical life span for an individual atom.

Exponential functions of decline also operate in the game "Selection." For the losing player, the final phase of the game follows an exponential law of decrease. Because of the limits the size of the board imposes on the game, exponential growth is possible only in the initial phases of the game and is the product of a streak of luck.

Another typical property of the exponential function is that it actually reaches the limiting values of zero (in the case of decrease*) or infinity (in the case of increase) only after an infinitely long period of time, however precipitous the process itself may be. This asymptotic behavior is directly correlated to the invariance of the half-life. Rapid growth simply means that the doubling periods are short.

But what happens when the doubling time is no longer constant but keeps shrinking with time? This is exactly what we are experiencing with the present population explosion, in which events occur with breakneck speed. If growth were unlimited, then the growth curve would reach infinity in a *finite* period of time. In this case, mathematicians speak of a singularity. By this they mean the point where a function, such as a hyperbola, disappears into infinity. This can, of course, never actually happen because, in our finite world, limitations automatically go into effect and prevent a quantity from becoming infinite. But the restrictions imposed by this process can have such catastrophic effects for the individual that his viability is threatened.

---

*If we kept cutting a continuous quantity in half, it would take an infinite number of cuts to arrive at zero.

11.2   EXPLOSIVE GROWTH

We shall now analyze various cases of exponential and hyperbolic growth with the help of a bead game.

In this game "Growth" the strategies that can be chosen are representative of certain growth mechanisms. It is, of course, impossible to simulate on a limited playing board an unlimited process of reproduction extending over many generations. Only in the initial phase of the game, when the bead density is still relatively low, are the conditions necessary for the simple laws of growth approximated. It is instructive to play this phase through attentively, although this takes some patience.

For a detailed analysis of the first strategy, it is best to start with only a single bead. This bead occupies one square and has a defined neighborhood consisting of the four orthogonally adjacent squares. We now have to hit one of these five squares in order to double this one bead. On a board of 64 squares, the chances of doing this are 5 in 64. In other words, we will need an average of 13 rolls to succeed.

Theoretically, we should have separate rolls of the dice for each of the two beads now on the board because the two reproduction processes are independent of each other. But in practice we speed up the pace of the game by letting one roll of the dice apply to both beads on the board, thereby utilizing the greater probability for a hit. As long as the board is large enough to prevent any appreciable overlapping of neighborhoods—and that is possible in actuality only on a very large board—the number of beads will keep doubling, on the average, after a given number of rolls, e.g., 13. That is characteristic of the exponential law of growth.

As the playing board becomes more densely occupied, its limited size brings about a clear deviation from the simple exponential law. Once there are enough beads on the board that all the remaining empty squares belong to a neighborhood, every roll of the dice represents a hit and inevitably introduces another bead onto the board. The speed of reproduction is maximal now, but it is constant, and the quantity therefore grows only in linear fashion.

Table 12. BEAD GAME ''GROWTH''

The game is designed for two to four players. It requires a board with coordinates and the appropriate dice. A board with 64 squares works best. Each player has an adequate supply of beads of one color.

At the beginning of play, the players take turns rolling the dice for a predetermined number of rounds, six, for instance. Each player places a bead on the square he has selected with the dice. If this square is already occupied by a bead of another color, the player has to pass. (This rule remains in effect throughout the game.) But if the square is occupied by one of his own beads, he may move this bead to any (strategically advantageous) square he chooses. (Neighboring beads are of some advantage in the further course of the game.)

At the end of the initial phase, each player has a maximum of six beads on the board. Now the game as such begins. Each player chooses a strategy that seems most advantageous to him in terms of the distribution of his beads, and he also announces this strategy to the other players. He is obliged to adhere to this strategy until he announces that he will change to the "next higher" one. The following strategies can be used:

1. The player whose turn it is may roll the dice once. If the dice select a square occupied by one of his beads or any one of four squares orthogonally adjacent to such a square (see Figure 46), he may place one more bead on a square to be selected by a roll of the dice.
2. The player may roll the dice twice in a row, regardless of the results of the first roll. If he selects a square with at least two of his own beads on neighboring squares, he can introduce two new beads into the game and place them on two squares to be selected by the dice. Neighboring regions are defined here to include the square on which a bead itself is located as well as the four orthogonally and the four diagonally adjacent squares (Conway neighborhood). The square selected by the dice has to lie in the overlapping area of two neighborhoods (see Figure 46).
3. The player may roll the dice three times in a row, but he has to hit a square that lies in the overlapping area of at least three neighboring regions of his own beads (see Figure 46). He may then introduce three new beads onto the board.

(Strategies 4 on up to the highest one, 9, follow the same pattern. In each case, the player may roll as often—and, if successful, introduce as many new beads onto the board—as the strategy he announced stipulates.)

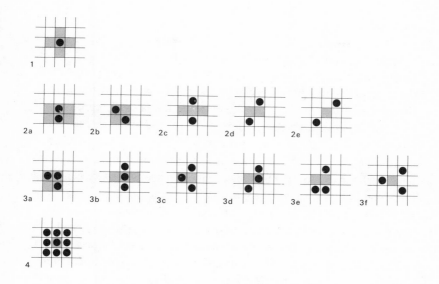

Figure 46. Definition of neighboring areas in the game "Growth."

1. For the one-bead strategy, only the von Neumann neighborhood applies. It includes the occupied central square as well as the four orthogonally adjacent squares (gray).

2. and 3. For strategies based on two or more beads, the Conway neighborhood applies. It includes both the diagonally and orthogonally adjacent squares. But the rules here apply only to those squares that are in the overlapping areas (gray) of the beads in question.

4. In the nine-bead strategy, a player has to hit the central square (gray) in a region occupied by nine beads.

The winning player is the one who has the most beads on the board when it is completely filled. The neighborhood rules of the second phase are not affected by the presence of opposing beads, except for the restriction that a player cannot place one of his beads on a square already occupied by an opposing bead. (In this version of the game, removing opposing beads from the board is not allowed.) If a player has new beads to place, he keeps rolling the dice until they select free squares for those beads.

Note. The cautious player will first choose the relatively safe first strategy. In an even distribution of beads, this strategy gives him the best chance for placing new beads on the board. But if the initial distribution produces configurations that make a higher strategy advantageous, it is advisable to adopt it early.

Similar limitations hold for the second and third strategies, too. In these strategies, the rate of increase is proportional to the square or the cube of the number of beads on the board. A reasonably correct simulation of the "higher" or superexponential laws of reproduction requires procedures similar to those used for the first strategy. Here, too, we have to assume a sufficiently thin "dilution" and a nucleus of at least two (or three) beads in a configuration making up a special neighborhood as required by the non-linear reproduction rule. Unless these conditions are met, the strategy in question cannot initiate the growth process. The probability of hitting such a nucleus is relatively small in the case of a chance distribution of two (or three) beads on 64 squares, but the probability increases as the game proceeds.*

This game illustrates the explosive nature of "superexponential" growth. If the higher strategies are adopted early in the game and maintained throughout it, the doubling times will become dramatically shorter, whereas the first strategy would maintain a nearly constant doubling time over long periods of play. The constant shrinking of doubling times is the most striking characteristic of hyperbolic growth. It is caused by a non-linear rate function. In other words, the rate of reproduction increases more rapidly than the quantity in question.

We could object at this point that the difference between the exponential and hyperbolic laws of growth is insignificant from a practical point of view. Since finite dimensions prevent the hyperbolic function from ever reaching its singularity anyhow, and since the curve of exponential growth will depend on whatever time constant we choose to apply, the law of hyperbolic growth per se is at best only of "theoretical" interest. In the next chapter we will see, however, that if certain limits are imposed on growth, the shrinking of the doubling times characteristic of the hyperbolic law will have significant qualitative and quantitative effects on the selective behavior of the system.

---

*To simulate a specific hyperbolic law of growth correctly, we would have to make a statistical redistribution of all the beads on the board for each round. Only in this way could we correctly re-create the probability distribution for configurations of pairs, triplets, etc.

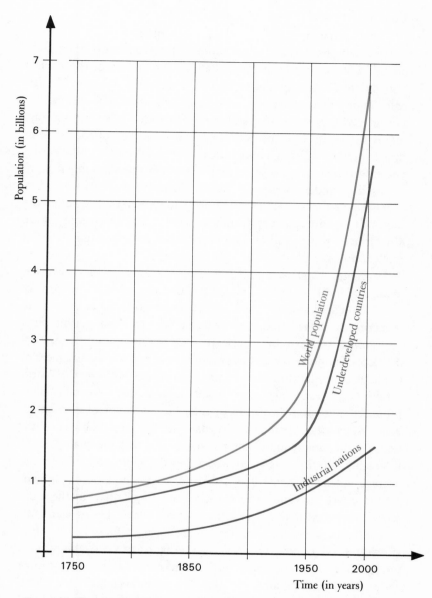

Figure 47. The growth curve of the world population (red) is broken down here into components representing the industrialized nations (green) and the underdeveloped countries (blue). The assignment of countries to these two categories is based on a table in Paul Demeny's article "The Population of the Underdeveloped Countries."[71]

How can changes in the conditions of reproduction be brought about? A great number of factors affects the growth of the world's population. The greater availability of medicines and improved hygienic conditions have enormously reduced illness and epidemics. Infant mortality is decreasing and the average life expectancy increasing. Rising birth rates and declining death rates will, of course, be in direct correlation only if rising life expectancy affects the childbearing years. Superexponential population growth is therefore much more dramatic in underdeveloped areas, as Figure 47 illustrates.

If life expectancy past the childbearing years increases, the doubling time will become constant again (see the age distributions in Figure 48). Growth would again become exponential, assuming that the birth rate exceeds the death rate.

As Figure 47 shows, global statistics can give a very distorted picture of local conditions, and stabilizing measures should always take such conditions into account. In India, for example, a drastic reduction in births is indicated, whereas in Argentina a rising birth rate would be desirable. Excessive population growth in underdeveloped countries has already resulted in significant disturbances of the natural equilibrium in many areas. Measures taken by the World Health Organization to reduce epidemics and infant mortality have produced this jump in the birth rate, but have not been counterbalanced by measures that would guarantee appropriate living conditions for the rapidly increasing masses. This is why the quality of life is steadily deteriorating in underdeveloped regions and has, on the average, already fallen below levels necessary for a decent human existence. At the same time we must not forget that, in the most wretched circumstances, a large number of children can give an individual family a certain advantage in the struggle for survival. Measures to alleviate poverty have to be designed to break this vicious circle. One thing that is certain is that external aid alone will not solve the problem. In this connection, the contrast between India and China should prove instructive to industrial nations intent on helping poorer nations. Ideas and ideals that can be realized are more significant than adherence to any specific ideologies.

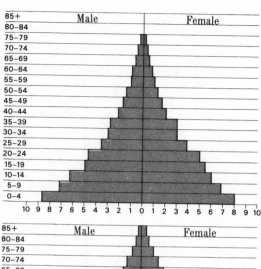

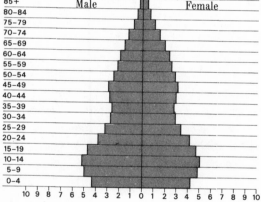

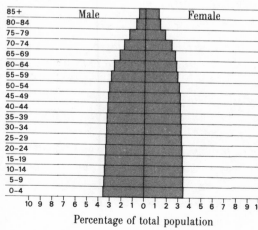

Percentage of total population

Figure 48. Age distributions within populations. The top pyramid represents the population of Costa Rica in 1950 and is characteristic for an underdeveloped country. The graph in the center shows the age distribution in the USA in 1970.[72] The narrowings in this chart reflect economic recessions. The bottom graph is typical of a stationary population in a country that has a high standard of living and has not experienced any "catastrophes" for at least two generations.

12

Limited Space and Resources

In a spatially limited environment, growth leads to saturation. There is a maximum limit for the total population. Individual subgroups, however, will display highly differentiated behavior in accordance with the particular law of growth affecting them:

1. *coexistence (with linear growth or mutual stabilization)*
2. *competition and selection (with exponential growth)*
3. *once-and-for-all decision (with hyperbolic growth)*

Where the exponential law is in effect, selectively advantageous variants can emerge in the saturation phase. The criteria for a variant to qualify are set down in "prohibiting clauses." But where the hyperbolic law is in operation, a selective decision, once made, is practically irreversible. On the playing board, we can simulate the effects that different laws of growth have in a saturation phase.

The evolution of life has passed through all phases of growth. Global restrictions would lead not only to ruthless competition but also to drastic limitations in the potential for further human development. Pressing as the problem of population control is, we must not forget that it should be solved regionally and not globally.

12.1   COEXISTENCE

In the last chapter we described bead games designed to illustrate the three fundamental laws of growth. In these games, we found that the limitations imposed by the size of the board had a decisive influence on growing populations. Patterns of growth soon deviate from their original tendencies and, in the saturation phase, act in accordance with completely new laws. This effect is by no means restricted to the playing board but is a routine occurrence in the real world. We will therefore analyze it in more detail here.

There are no unlimited spaces in our world. For the foreseeable future, at any rate, the limits of our planet represent the limits of our living space. If the population of the earth continues to grow at the present rate, there will be only one square yard of living space left for each human being in five or six hundred years.

Extrapolations of this kind, of course, however justified they may be from a normative point of view, hardly offer accurate predictions of what will actually happen; and a futurology that deals solely in extrapolations is of little value to us. The recent oil crisis made this point vividly clear, calling generally accepted predictions like "Japan will be the greatest economic power of the future" into question. On the other hand, extrapolations can be useful for anticipating trends and giving early warning of catastrophes in the making. But quantitative predictions are of value only if margins of error are defined at the same time. Any prediction for a stationary autocatalytic system is limited in time. Indeed, we should always mistrust such predictions unless they also state when the predicted value will become meaningless because of the growing margins of error.

Restrictions on growth in a system made up of differentiated, self-reproducing individuals will evoke competitive behavior among these individuals, even if they are of basically amicable disposition. The rules Conway devised for his "Life" game, for instance, do not contain any combative elements; but in spite of this, they create, as we have seen, a whole arsenal of weapons. As the history of mankind shows, if there are weapons available, they will be used. Limitations on living space inevitably lead to infringements of other people's rights, even though there is no harmful intent. Someone who lights a cigarette in a crowded subway train does not do so with the intention of polluting the air for others. One individual too many in a fully occupied space automatically limits the freedom of the others and represents the first step toward aggression. Animals will defend their territory to the death, and human beings too are unsuited by nature for living in an overpopulated world. Overcrowding will elicit aggressive behavior from them as well.

If there are limits imposed on quantity, the different laws of growth will affect the representation of individuals differently. We might be tempted to think that it would be relatively unimportant which law of growth was in operation. At first glance, all that would seem of importance is "how fast," not "how," the competing individuals multiply. This is correct to the extent that certain classes of individuals, provided they multiply more rapidly than their competitors, cannot be displaced by them unless some interfering interaction is at work. But in spite of this, the three laws of growth will produce totally different qualitative results in the representation of individual species. This may not seem surprising for linear and exponential growth because they are based on different strategies of growth. It is worth noting again here that linear growth results from a constant growth rate, i.e., one that is independent of the quantity in question. In the pay-off matrix on page 28, we have designated the strategy at work here as "indifferent" (S_0). Exponential growth, on the other hand, results from a rate function that is linearly dependent on the quantity in question. This represents a special instance of the conforming strategy.

There is no genuine competition involved in linear growth, no matter what shape limitation takes. The constant rate of increase is simply compensated by the rate of decrease, in which the individuals, in accordance with a natural law of decay, disappear in proportion to their presence. This process results in a stable distribution (state $+0$ in the matrix on page 28), and the proportions in the population remain constant. But if the rate of decrease is constant as well, i.e., independent of the number of individuals, the system is indifferent

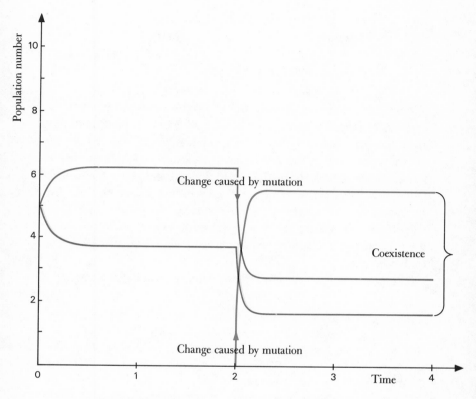

Figure 49. The coexistence of two species (blue and green) results from a limitation of linear growth. In the example shown here the ratio of the growth rates blue : green is 5 : 3. Since the sum of the initially equal subgroups is kept constant, green has to decrease by the same amount that blue increases. This process is completed once both subgroups have arrived at the ratio reflected in their growth rates. Under these conditions, a new mutant (red), regardless of its efficiency, can easily make rapid advances. The ratio between all the species present will settle into a pattern reflecting their rates of growth.

(00) and drifts randomly through all possible population numbers. In neither case can a clear selection occur. The result is a coexistence of all differentiated systems of individuals, and these systems can be represented in all kinds of proportions (see Figure 49). Among self-reproducing individuals under the pressure of selection, coexistence always requires a special stabilizing interaction that we will discuss later on.

12.2 COMPETITION

Global limitations on growth would produce totally different results for exponential and hyperbolic growth (see Figures 50 and 51).
Here, proportionality between the growth rate and the quantity in
question produces a clear selection. A coexistence of different populations is impossible once the saturation point has been reached.
Our fourth bead game, "Once and for All," illustrated this clearly.
Exponential laws of growth are responsible for selection in Darwin's sense. In reality, of course, selection occurs in a variety of
ways because the limiting conditions affecting it are not clear-cut
as a rule.

The premises for clear-cut selective competitive behavior are:

1. The individuals are made up of the same material, i.e., they are, at least
   indirectly, dependent on the same sources of food.
2. Limitation imposes stationary behavior on the total population. That
   means that the sum of all competitors remains constant. One class can
   grow only at the cost of another.
3. There are no stabilizing interactions at work between species.

These premises make clear how a variety of species could develop
in the biosphere although all were self-reproducing and none was
characterized by absolute stability. Perhaps the third premise requires some further explanation. A stabilizing interaction can, for
example, take the form of one self-reproducing species supporting
another in its reproductive process. But this kind of interaction by no
means guarantees the existence of the other species. If the second species dies out—perhaps because of an unfavorable fluctuation—then
it cannot arise again, because the template for its reproduction no
longer exists.

The situation is different in a reaction cycle in which an individual
species is not self-reproducing. The essential point here is cyclical
continuity. A catalyzes the formation of B, B that of C, and so on until
some product X finally helps form A again.

If this cycle includes enough partners, there will be no fluctuation catastrophe unless all the partners die out at the same time. The nucleic acids multiply in a complementary cyclical pattern. Each chain catalyzes the production of a complementary copy of itself, i.e., a negative, that in turn triggers the formation of a positive. Seen as a whole, a cycle of this kind is self-reproducing. In addition, the partners interact to stabilize each other.

In connection with the premises for selection, we must add a note on the concept of global limitations. In a system made up of different kinds of individuals that may be distributed in separate regions, we can, on the one hand, place limits on certain kinds of individuals or on certain regions by direct selection. On the other, we can employ global controls. That means that we can hold the number of *all* individuals constant and leave it up to internal conditions in the system, that is, to the individuals in question, to adjust themselves to the overall limitations.

The first Club of Rome report by Dennis and Donella Meadows[73] was written exclusively in terms of global limitations. Justified criticism of the recommendations in this report focused on the extremely disturbing consequences of global restrictions. Without a global government or globally binding treaties, such restrictions would produce nothing but chaos. Global restriction, without further qualification, would lead to a ruthless competition that might well end with total self-destruction of the entire system. We will discuss this issue in more detail in the next chapter. Here we will merely point out that selective behavior varies greatly, depending on how and what we limit. We can limit the supply of energy as well as rates of reproduction, or we can limit populations themselves. Because of the enormous differences in natural conditions in different parts of the world, we have to consider carefully whether we should stabilize population

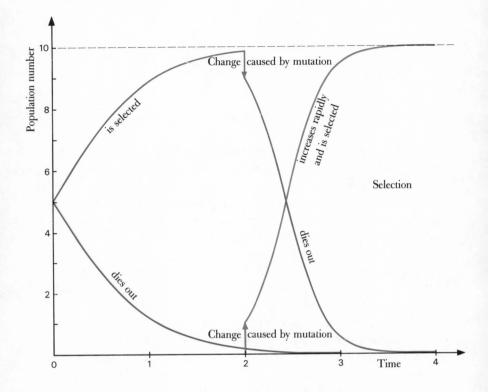

Figure 50. Selection will result from the competition of two species (blue and green) provided that an exponential law of growth is at work. The ratio between the parameters of the speed of growth (initial rates) is again 5:3 for blue and green. In this case, there is no coexistence; the less viable component (green) dies out. A mutant (red), arising at the cost of blue, can make rapid advances if it has a clear selective advantage. In the case shown in the graph, the initial growth rate for red is twice as high as that of blue. Given these conditions, red increases rapidly and completely displaces blue.

numbers or whether we should adjust them to suit the production of food and energy. In underdeveloped areas, we can have only a slight influence on production. In such a case, it would be advisable to lower population numbers to suit local conditions. In industrial nations,

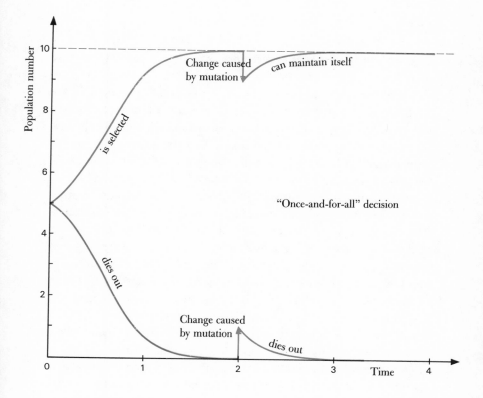

Figure 51. A "Once-and-for-All" selective decision results from competition under conditions of hyperbolic growth. The ratio between blue's and green's initial growth rates is 5:3. As in the example in Figure 50, selection occurs here, too, but in a much more abrupt way. The decision here cannot be reversed once the selected species is present in sufficiently large numbers. A mutant (red), even if it has a basic growth rate that is twice as high as blue's, cannot establish itself. Red's selective advantage is only a theoretical one in this case because it could be of practical value only if the populations of blue and red were nearly equal. To displace blue, red's single copy would have to have a growth rate that could balance out the population ratio of blue and red. If this is not the case, red will always die out.

however, population numbers can be held more or less constant and production rates of consumer goods can be adjusted and regulated in terms of those numbers. Population growth, production of goods, emission of pollutants, and capital investment all behave in accordance with their own laws and require regulatory measures that take these laws into account.

We will now turn to our third case, a non-linear rate function. We might not expect that the results of exponential or hyperbolic growth in a limited system would vary drastically from each other. Both these laws of growth are based on the conforming strategy S_+ (see p. 26). Seen mathematically, a hyperbolic function, unlike an exponential one, displays singularities—points at which infinity is reached in finite time. But a limited quantity cannot, after all, become infinite. In other words, the effects of limitation will be felt long before those characteristic of a given singularity can ever set in. We therefore never have a chance to observe this crucial difference between exponential and hyperbolic growth. The exponential multiplication of neutrons in uranium is, for example, adequate to produce the massive explosion of an atomic bomb. What could be added by an increase of neutrons that is greater than exponential? The exponential law of cell multiplication is also adequate to induce selection in a limited living space. What more could hyperbolic reproduction accomplish?

There is, in fact, a very important difference between exponential and hyperbolic growth. The end result of hyperbolic growth is a final selection (see Figure 51). By contrast, exponential growth in a limited environment will result in the reproduction of the *one* component that is the fittest, *relative to the competitors present in the distribution.* But if a new and better-adapted variant appears, it can establish a foothold and displace its predecessor, provided its selective advantage is great enough (see Figure 50).

If the pressure for selection is moderate, as it generally is in nature, exponential growth creates conditions that allow several variants to tolerate each other. This explains the wealth of variety that we observe in biology. Once the phase of cell individualization was completed, reproduction in evolution essentially followed an exponential law. If the conditions of growth had been controlled by a hyperbolic law, we would not have the marvelous variety, the spectacular "symphony of animate nature" that we do.

By means of a game, we can illustrate the consequences of global limitation under conditions of competitive growth.

Table 13. B E A D   G A M E   ''L I M I T A T I O N   O F G R O W T H ''

The requirements of the game are a playing board with 64 squares, 64 black beads, 64 white beads, and a pair of octahedral dice. This game is designed for only one player. It is best to play the different versions one right after the other and, if possible, several times. In this way, the player gains an understanding of the effects of limiting growth on a global scale.

Version 1. 62 squares are filled with black beads, one square with a white bead, and one square is left empty. The dice are rolled alternately for birth and death, and the beads are shifted according to the conforming strategy S_+. In other words, every bead selected for birth is doubled, and every bead selected for death is removed from the board. (If the empty square happens to be selected, the dice are rolled again.) The strict alternation of rolls of the dice for birth and death maintains a constant population saturation. The game is over when one of the two colors of beads dies out.

Version 2. We proceed here as in Version 1 but with one difference. If a white bead is selected by a roll for death, the player must then make a yes-or-no decision. He might do this by tossing a coin, for example. Only if the decision is yes will the bead be removed from the board. This additional rule can be varied by using a normal die to make the yes-or-no decision. In this case, the probability for yes is no longer ½ but can range from 1/6 to 5/6. With a death probability of 5/6, the white bead will be removed from the board if any number from one to five is rolled but will remain in place only if a six is rolled.

Version 3. The rules of Version 2 apply here, too, but we start with a different distribution of beads on the board. If the probability for death for a white bead is ½, the player begins with 2 white beads and 61 black on the board. If the probability for a yes decision is 5/6, the player should begin with more white beads on the board. It is instructive to play this version starting with 2, 5, and 8 white beads on the board.

Version 4. The player places 6 white and 48 black beads on the board. This leaves 10 empty squares. Now he rolls the dice and proceeds according to the following rules: For birth, the dice are cast twice in a row. But new beads can be placed only after the second roll and only if the dice have selected the same color both times. (If the dice select empty squares, the roll can be repeated.) If white is selected two times in a row, the player places ten white beads on the board at once. But if black is selected twice, he places only one black bead on the board. Death is governed by the rules applying in the preceding versions: Whatever bead the dice select, black or white, is removed. In this version, of course, the strict alternation of birth and death cannot be maintained. The rolls of the dice have to be regulated in such a way that there are always 10 empty squares available for an increase of 10 white beads.

We should warn the reader that this game has a purely didactic function and does not hold many surprises. Versions 1, 2, and 3 are, obviously enough, based on an exponential law of growth. Only Version 4 makes use of a hyperbolic law. Rates of formation (doubling) and disintegration (removal) for both black and white are directly proportional to distribution density.

In Version 1, every bead has the same probability of being doubled or removed. In Version 2, however, the individual probability for disintegration (removal) is smaller for white than it is for black. In addition, the ratio of probabilities for doubling and removal is twice as great for any white bead as it is for any black one. But despite this, white will almost always lose in Version 1 and will lose about half the games in Version 2. White is likely to win only in Version 4. We can observe an interesting relation here if, as suggested in the description, we vary white's proportions of probability for doubling and removal as well as its initial number of beads. We will see that white will have a better than 50 percent chance of winning only if its initial number of beads stands in a certain relation to the ratio of the individual rates of doubling and removal.

For example: The probability for doubling for any white or black square is $1/64$. White's probability for removal is $5/6 \times 1/64$; black's is $1/64$. Therefore, we have to have about six white beads on the board at the beginning of play so that white will have about a 60 percent chance of winning.

This example shows, for one thing, that exponential growth can allow a new component, provided the component is fit enough, to establish itself and displace its predecessor at any time. In addition, we find that a dominant population is protected against minorities by a qualification clause. If the selective advantage of a new mutant is only one hundredth of the selective value of the existing wild type, the mutant would have to appear in 100 copies to have a better than 50 percent chance of establishing itself. If the relative advantage is 10 percent, only 10 copies would be needed to produce the same effect. If the advantage were 100 percent (which means that the fitness parameter of the advantageous kind is twice as high as that of the less advantageous kind) a single mutant would be able to establish itself in the majority of cases.

The new variant must have clear advantages if it is to stand up to competition successfully. This phenomenon is reminiscent of the prohibiting clauses that parliaments use to protect themselves against splinter groups that represent only minute segments of the electorate. By making use of a similar ruling, evolution has avoided many detours and dead ends. It would be very uneconomical, for the sake of only a minor advantage, to dissolve and completely replace an already existing system. Still, it is always possible for a mutant with very little advantage or none at all to establish itself by a random drift process. The slighter the selective pressure—that is, the more often increase beyond a stationary level is locally permitted—the more often this is likely to occur.

12.3 "DECISION"

Now we will consider the fourth version, the one based on a hyperbolic law of growth. The chance that two successive rolls of the dice will select the same color bead is proportional to the square of the probability of selecting it with only one roll. The probability of selecting it with a single roll depends on the relative density of that color on the board. The periods required for beads to double themselves will become constantly shorter as the number of beads increases.

To demonstrate this principle, we will take a concrete example. If there are 6 white beads and 48 black ones on the board, the probability of selecting the same color twice in succession is $(6/54)^2$ for white and

$(48/54)^2$ for black. (The denominator is 54 rather than 64 because the empty squares do not count.) Despite the fact that white beads increase by 10 if they are selected twice in succession while black increases by only one bead at a time, it is still practically impossible for the more efficient white beads to establish themselves unless the player is lucky early in the game and can introduce 10 new white beads onto the board at that stage. In that case, of course, white will increase more than it will decrease, and this trend will grow more pronounced until black is completely displaced. We have deliberately chosen the numbers in this example so that the one result will occur as a rule but the other is not precluded.

The results in this game would be much more striking if we used a considerably larger number of beads, e.g., a thousand or even a billion. With such numbers, there would be a clearly defined threshold in the initial distribution that would inevitably produce a "once-and-for-all" decision, and the species emerging as the victor would not tolerate a rival for the remaining course of the game. This is a typical consequence of the hyperbolic law of growth. The rate of increase for each individual species is not simply a constant that is characteristic of that species, as the doubling period is in exponential growth, but it also contains as a factor the number of already existing similar species because they function as reacting partners of the original species in question. Once a population is established, i.e., once its number has risen to a million or a billion or higher (the smallest number of macromolecules that could be observed experimentally), a mutant occurring in a single copy would have no chance against such overpowering numbers. Its rate of increase in the initial stage would be much too small no matter what advantages it might have.

The situation is very different with exponential mechanisms of growth. Here the advantage has to overcome only a relatively weak statistical prohibiting clause (see pp. 226–7) to be able to establish itself. This prohibiting clause deals in percentages; the one applicable to hyperbolic growth deals in multiple powers of 10.

In the evolution of species, simple cell division and reproduction by sexual combination have produced laws of growth that are essentially exponential. The hyperbolic phase of the present population explosion cannot be ascribed to an inherently non-linear rate function. It is caused instead by a rise in average life expectancy. This means that more human beings reach reproductive age. This hyperbolic phase represents a transitional stage.

In the early stages of the evolution of life, when the molecular self-organization of systems capable of reproduction took place, there were surely mechanisms inherent in the overall system that triggered a phase of hyperbolic growth. The division of labor between the nucleic acids and the proteins, a division resembling that between a legislative and an executive, must have resulted from a non-linear law of reproduction. The rate of increase in a primal system depends, on the one hand, on the presence of nucleic acids that contain the information for the construction of the system. On the other hand, however, it is proportional to the amount of the functional proteins, an amount that is in turn correlated to the amount of the nucleic acids responsible for coding those proteins. The combination of the two molecular systems into a "living," self-reproducing individual, such as a primordial cell, did not necessarily have to take place in one step but must at some time have been made permanent by a superimposed cyclical linkage of the molecular partners. Every nucleic acid represents a small self-reproducing cycle (positive$\rightleftharpoons$negative). In the linked system, many such reproductive cycles (the predecessors of genes) must cooperate via their translation products, the proteins. These, in turn, make use of their catalytic abilities to bring about coupling. We can say in general that a cooperative totality, as represented by such a hypercycle, can take shape only if the superimposed linkage itself forms a closed loop, i.e., if the end of the process ties in again with the beginning.

Two effects follow:

1. The individuals linked together in this way are dependent on each other and therefore coexistent rather than competitive. In the context of the whole cycle, each individual member is stable.
2. The cycle as a whole is extremely competitive in its interaction with other systems, and it evokes irreversible "once-and-for-all" decisions.

The characteristics of such a system can be exemplified in a game.

Table 14. B E A D   G A M E   "H Y P E R C Y C L E"

A hypercycle is a cyclical linking of self-reproducing individual cycles. The individual cycles

and so on are combined into a new organizational form by a superimposed cyclical closed coupling:

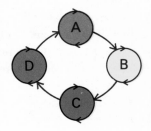

Competition between the individual cycles is transformed into cooperation by the coupling. Because of the non-linear, autocatalytic rate of increase, "once-and-for-all" selection among various hypercycles will take place.

Our game is based on a simple hypercycle made up of four components. This hypercycle can be represented by the following abstract reaction scheme:

$$A + B + M \rightarrow A + 2B$$
$$B + C + M \rightarrow B + 2C$$
$$C + D + M \rightarrow C + 2D$$
$$D + A + M \rightarrow D + 2A$$

A = red, B = yellow, C = green, D = blue. In our game, M plays no role but we have listed it here to represent the material from which A, B, C, and D arise in actual autocatalytic reactions.

The game calls for a playing board with 64 squares, octahedral dice, and a sufficiently large supply of beads in four different colors. At the beginning of the game, 16 beads of each color are placed on the board, filling it completely. Now the dice are rolled alternately for removal and doubling. The bead selected by the first roll of the dice is removed. Then the dice are rolled again. The bead selected by this roll can be doubled if, and only if, one of the four orthogonally adjacent squares is occupied by a color preceding it in the following cyclical order: red (A) precedes yellow (B), yellow (B) precedes green (C), green (C) precedes blue (D), and blue (D) precedes red (A).

The new bead, which has the same color as the bead selected by the dice, is placed on the square that was cleared by the preceding roll of the dice. These two steps are repeated alternately.

This game somewhat resembles "Selection," but because of the superimposed coupling, the game does not lead systematically to the selection of one color. But there is still a possibility that an extreme fluctuation can wipe out one color. The number of beads will decrease steadily at the beginning of the game despite the strict alternation of rolls for removal and doubling. This can be prevented if a removal is permitted only after a doubling has actually been achieved. The coordinates have only an indirect bearing on play in this model. The fact that beads introduced by doubling occupy squares emptied in an unrelated process reflects the constant mixing of reaction partners in nature. This game depicts the "rise and fall" of individual colors. Periodic fluctuations are typical for cycles of this kind, and in the oscillation of the bead distribution on the playing board all the colors will rise and fall in cyclical sequence (see Figure 52).

The following variation on the game depicts competition between two hypercycles. Four colors are arranged to form two independent hypercycles:

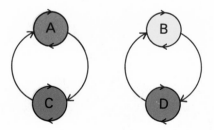

Two players place their beads alternately on the board. One player has red (A) and green (C); the other has yellow (B) and blue (D). Neighborhoods are defined here in terms of the complementary colors. The players take turns rolling the dice for removal and doubling of their own beads. This means that a player can make a move only if he rolls a square containing a bead of one of his colors. Otherwise, he has to pass. If he is able to double one of his beads, he may also remove one of his opponent's beads. The game ends when one of the players is unable to make any further moves.

This version of the game provides a further example of competitive decisions in hyperbolic growth. Oscillations do not occur in hypercycles with two components.

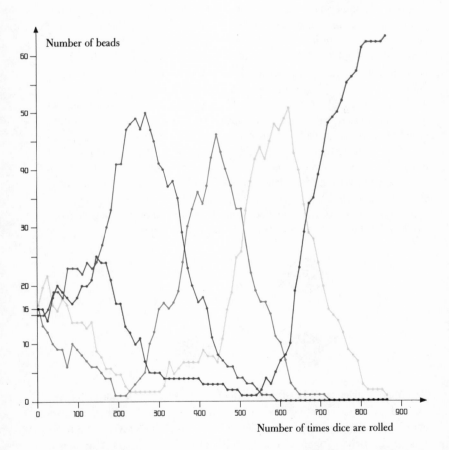

Figure 52. Computer graph of a "Hypercycle" game. The diagram shows how the individual populations of A (red), B (yellow), C (green), and D (blue) changed as the dice were rolled. A fluctuation catastrophe broke off this cycle.

We let a computer play the first version of this game over an extended period of time. The results are shown in Figure 52. Note the clear oscillations of the four colors and the collapse of the cycle as a result of a fluctuation catastrophe. This cycle will break down if even one component dies out. This component cannot renew itself because it cannot reproduce without a template.

A great variety of hypercycles are conceivable. Peter Schuster of the University of Vienna has classified them in terms of their mathematical properties (e.g., type of non-linearity, extent of linkage between components, etc.). The conditions necessary for the rise of hypercycles were obviously present in the prebiotic phase of evolution. Whatever individual cycle set the historical process of life in motion must have been based on a mechanism employing a non-linear rate of self-reproduction. A simple linear rate, characteristic of exponential growth, could not have created an interaction between legislative and executive, and it is this interaction that gave rise to the genetic code. This code must therefore have originated in a "once-and-for-all" decision. All organisms, from coli bacteria to human beings, utilize the same universal code and the protein machinery that goes with it. The fact of a universal genetic code had first prompted the conclusion (a) that all life has a common origin and (b) that this original event was so rare that it happened only once (cf. our discussion on the origin of chirality in Chapter 7, pages 121–4).

The first conclusion is no doubt correct. If several original events had initiated independent and successful developments, we should now find different code and structure patterns among living beings.

The second conclusion is, however, by no means convincing, and it is very likely false. It would be remarkable indeed if all the individual events that were necessary for the development of life, unique as they are, happened only once and at exactly the right moment. It is far more likely, and we have reason to assume, that they all happened relatively frequently and consequently admitted the possibility of many alternative combinations. This is how, in later phases, the great variety of species originated. The historical uniqueness of code and chirality would not then represent evidence that life arose in a single event. It would be instead the consequence of a "once-and-for-all" decision among a large number of alternatives, a decision that was made once and for all in the hyperbolic phase of growth.

Games like "Hypercycle" are being played in many laboratories today, but instead of beads, scientists are using molecules. Nucleic acids act as storehouses for molecular information, while replicating and degrading enzymes are used as molecular sewing and cutting machines. Bacteria and viruses can be isolated and—as we will show in Chapter 13—put back into operation in a "bioreactor" under controlled physical conditions. Here men replay nature's game. The rules are the same as in our bead games, but the actual playing of this "game of life" requires a vast amount of detailed knowledge, skill, and imagination on the part of the scientist.

We can summarize as follows the results of our bead games as they apply to global limitation in conjunction with specific laws of growth:

1. Linear growth always leads to coexistence and to population densities that are, on the average, determined by the ratio of the rates of formation and decomposition.
2. Exponential and hyperbolic growth result in a clear selection of one species unless stabilizing interactions among different species enforce their coexistence.
3. In the case of exponential growth, "qualified" competitors (i.e., mutants with a clearly defined selective advantage) can establish themselves at any time. In hyperbolic growth this is practically impossible once a species has qualified and established itself.
4. Rules 2 and 3 apply consistently only if there are no functional links between the competitors. Links of this kind can lead to a mutual stabilization of the partners involved or to a stiffening of competition between them or even to a total extinction of them all.

The superexponential (hyperbolic) increase in population that we are experiencing in some parts of the world today is the result of influences that are effective only at a certain stage of growth. This stage will end once the age charts for all regions of the world reflect a cylindrical distribution (see Figure 48). Whether this situation will produce peaceful coexistence, harsher competition, or global catastrophes will depend largely on natural constraints, such as the availability of raw materials and energy, as well as on ideas for the ecologically sound use of available resources. Above all, it will depend on measures men devise in the light of these conditions.

13

From Ecosystem to Industrial Society

An economic system, like a natural ecosystem, is influenced by a number of parameters. On the basis of game theory, we can establish relationships that promise optimal results. These relationships have counterparts in non-equilibrium thermodynamics, which may, in turn, be used to provide the foundations for a quantitative, analytical economics.

Dynamic systems are regulated essentially by fluxes and by forces or potentials. There are subtle methods of regulating fluxes or potentials to control growth in finite systems. These methods can be adapted to the specific nature of the productive processes in order to preclude instabilities or catastrophes.

13.1 ANALYTICAL ECONOMICS

Games like "Life," Chess, and Go illustrate how easily complex situations can arise from the iterative combination of the simplest principles. But economists deal with reality. They have to determine to what extent basic mechanisms that are generally accepted as valid and are known to function under precisely defined constraints can be used to analyze, and perhaps even influence, the facts they encounter in reality. There are impressive examples of how simple laws have been used to analyze highly complex economic and social phenomena.

In his acceptance speech for the Nobel Prize,[74] the economist Paul A. Samuelson of M.I.T. outlined the possible uses of a quantitative analytical economics; and, drawing on the example of the Le Châtelier principle, he stressed the parallels between optimality criteria in economics and equilibrium relations in thermodynamics.

Le Châtelier's principle is a generally valid but, unfortunately, an often misinterpreted principle of thermodynamics. It states that a system in chemical equilibrium, when subjected to the stress of a change in temperature, pressure, or concentration of components, will always shift the equilibrium in the direction of minimal stress, i.e., in the direction that opposes, nullifies, or uses up the stress exerted by the change. In its exact formulation, as developed by Carl Wagner and Max Planck, Le Châtelier's principle gives generally valid information on how a physical system will react to a change in external stress.

If air is compressed in a bicycle pump, not only is the air reduced in volume but its temperature is also raised. Le Châtelier's principle is not so much concerned with the obvious consequence that the volume of air is reduced but rather with the fact that this reduction is, in absolute terms, greater if the air is compressed slowly enough to allow an exchange of heat with the environment than it will be if the pump is insulated to prevent the loss of heat and if the temperature of the air rises accordingly. It is not so easy to grasp this principle intuitively. It may be of some help if we consider that a quantity of gas trapped in a certain space will exert greater pressure on the walls of the container if its temperature is increased. If we raise the temperature of the gas as we compress it, then we will not need as great a reduction in volume to arrive at a given pressure as we would need if we compressed the gas without heating it.

But what does this thermodynamic principle have to do with economics? Samuelson learned that analogous relations between forces and their effects also apply in economics.

Let us look at an example. Wages and the amount of time worked or the price of goods and the amount of goods stand in similar relations to each other as variables of force and extent in thermodynamics, e.g., pressure and volume or temperature and entropy. In a free economy, a rise in wages or prices will always reduce consumption. If, for example, the hourly wage for domestic help rises from two to three dollars, the employer—responding to economic laws and trying to make optimal use of his financial means, assumed to be constant—will not be able to afford as much domestic help in the future. That is obvious.

But now let us consider two different conditions that could be imposed on the system:

1. Despite rising wages, all prices for consumer goods remain constant.
2. The supply of goods available to the buyer is rationed; or the buyer sees no reason, because of a general rise in wages and prices, to change his *relative* consumer habits.

Many of us have experienced rationing during wartime. Similar conditions still exist in many countries; people are frugal in buying necessities but are unable to spend their money on other things simply because the goods are not available. They are therefore forced to maintain consumption at a constant level.

If we consider the above example of domestic help in the context of the first condition, it is clear that an employer would cut back on his domestic help more drastically here than he would under the second condition. He would do this so that he could spend a relatively larger part of his budget on consumer goods, whose prices remained constant. In some cases the employer may not restrict his (absolute) consumption of goods at all and will make up for the rise in wages completely by reducing the amount of help or, perhaps, by doing without it altogether. This will hold true only if two conditions are met:

- All other parameters that may be affected by changes in wages are controlled, i.e., they are kept constant. This is the assumption of a closed system in thermodynamics.

- Both before and after the rise in wages, personal expenditures are optimally distributed in terms of one's own values. This corresponds to achieved equilibrium in thermodynamics.

Just as thermodynamic relations between pressure and volume or temperature and entropy are regulated individually for any material by the intrinsic compressibility or specific temperature of that material, each one of us will establish criteria for how he spends his money according to personal taste. What is important is to stick to one's norms and not keep changing them arbitrarily. Otherwise, valid predictions are impossible.

In economics, as in a system of complex chemical composition, we have to deal with a number of such paired variables that are linked together in a great number of ways. We see an example of this in the relationships between raw materials, finished products, wages, and prices that a large company has to work out. The condition of minimal free energy required in the achievement of thermodynamic equilibrium (cf. Chapter 8) corresponds to the demand for maximum turnover and return. This maximum can be found by applying game theory (cf. Chapter 2). If a large number of paired variables affects how profitable an operation is, the problem can hardly be solved by intuition. The only things that count here are the facts and the exact correlations that clearly mark the path of minimal resistance. Samuelson's and other economists' approaches, along with game theory, have transformed economics into an analytical science whose successes are beyond debate. We must emphasize again, of course, that these laws cannot be expected to hold unless the necessary preconditions are fulfilled to the letter or at least have a dominant influence on the system. This kind of controlled situation is easy to achieve in a physical experiment, but the economist who is asked to make predictions about business trends often has to work from assumptions that are only partially valid.

The situation is more complex still when time-dependent variables and therefore non-equilibrium states are involved. In reality we hardly ever encounter reversible relations like those on which equilibrium thermodynamics is based. In business, too, it is not so much the absolute quantity of goods that counts but their rates of production and sale. The relation of cash flow to capital is a key factor in business considerations. Cash flow is, as its name suggests, a measure of cash movement made up of profits, depreciation, and retained earnings within a given period of time. Outside forces determine not merely changes that bring about static conditions but also changes that are dynamic and initiate further changes in the course of time.

13.2   FLUXES AND FORCES

Today's physicists use Le Châtelier's principle to characterize the various interactions between fluxes and forces in a stationary state. Reinhard Schlögl and Ilya Prigogine worked out the exact formulation of this principle in its generalized form.

We do not need to go into the abstract formulation of the principle, but can simply state that it postulates an order. It indicates, first, whether a flux associated with a force responds positively or negatively to a change in that force and, second, how the relative magnitude of this effect depends on the remaining parameters of force and flux.

Limitations can be imposed on a dynamic system by controlling the force or by keeping the flux constant. If the relationship between force and flux is variable, these two methods can have very different effects. One form of limitation cannot be equated with another. We have to decide carefully which is the appropriate method.

The influence that controlling force or flux can have on the limitation of growth can be studied experimentally in the limited ecosystem of the evolution reactor. In the reactor, which is illustrated schematically in Figure 53, a competition between different species takes place. In this case, the species are nucleic acids with alternative sequences that serve, in a cell-free medium, as templates for their own reproduction (cf. p. 275). For this process they need a molecular machinery made up of enzymes and regulating factors. These factors are supplied either by a constant flow of native material, that is, they are part of the given environment; or they are themselves, as

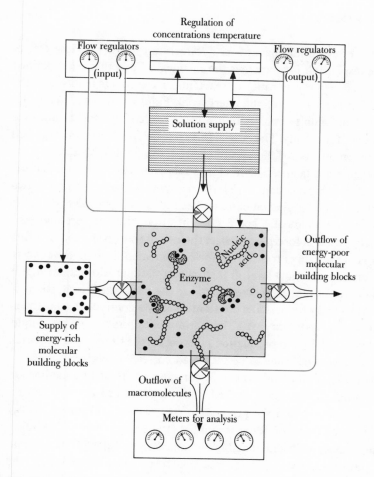

Figure 53. Diagram of an evolution experiment. Controlled amounts of energy-rich molecular building (●) blocks (nucleoside triphosphates of the building blocks A, U, G, and C) and of a solution that contains the required buffers, salts, and other factors flow into the reactor. Synthesized macromolecules (oooooo = RNA) and energy-poor waste products (o) (nucleoside monophosphates and pyrophosphate) flow out of the reactor. In addition, the system contains an enzyme (replicase) in a precisely regulated concentration. This enzyme joins the building blocks of nucleic acid together into chains. The building material is radioactively marked to permit constant control and analysis of the molecular composition of the contents of the reactor. The evolution machine can be operated either with a constant flow of energy-rich building material or with a constant concentration.

factors of evolution, part of the production cycle. The parameter of force responsible for growth is the chemical affinity that is determined by the concentration ratios of reactants and products involved in the transformation, i.e., of the energy-rich building material and polymers. This affinity can therefore be regulated by concentration or dilution of reactants. The flow of this material into the reactor (as well as the removal of reaction products from the reactor) thus represents the regulating factor. This factor is relatively easy to manipulate.

In regulating a system with limited growth we have to consider two limiting cases. In the first case, we keep the fluxes constant. By doing this, we automatically get a concentration level of reactants and thereby reaction forces that reflect the rate of transformation. In other words, if the rate of transformation in the reactor is high, the concentration sinks to a low level. If the transformation is slowed down, this level rises. In the second case, we keep the integral concentrations of reactants and products at a constant level by a continuous adjustment of fluxes in accordance with the transformation rate. If, for example, mutants with high reproduction rates occur, the supply of energy-rich building materials has to be increased correspondingly.

This model illustrates the different effects of these two methods of regulation. If the flux is limited, the energy needed for turnover remains constant, and the production rate adjusts itself to the energy level. If quantity or force is limited, the energy supply has to be continuously adjusted. The higher the catalytic rates of transformation are (assuming that the reaction force remains constant), the greater the increase of energy required will be. This kind of regulation presupposes a practically unlimited supply of energy-rich building material (nourishment), but it has the advantage of being less susceptible to crises. Keeping fluxes constant can easily lead to large fluctuations if autocatalytically reproductive species are involved. To maintain the viability of a hypercycle, for instance, a certain level of concentration is necessary (see p. 233). If there were large fluctuations in the concentration, individual partners could die out. The entire cycle would then break down.

In the plant and animal realm, growth is accompanied by multifariously interwoven processes of ecological regulation. These processes have to take highly complex constraints into account, conditions under which limiting cases of constant force and constant flows occur only rarely. This is also true of all the natural growth processes we observe in economies and societies. The moment we attempt to take regulatory measures we are faced with the question of how extensive an effect our limitation of force or flux will have. We have to consider carefully the conditions applying to each individual case to determine what the most appropriate form of regulation is. Limitation of growth cannot be allowed to threaten the stability of the system.

The difference between the control of flux and the control of extent of force can also be observed in the game "Selection." The beads have "self-reproductive" properties: a bead selected by the dice is doubled or removed in alternation. The rule that makes the doubling rate depend solely on the number of beads of one color present at any given moment and not on the availability of some building material is reasonable only if the supply of this building material is inexhaustible. The alternation of doubling and removal provides for a constant total number of beads in play. Version 3 of this game illustrates the effects produced by variations in the rate of flux: the higher the "value" of the bead introduced onto the board, the more often the dice will have to be rolled in the removal phase to reestablish balance. The game simulates an evolution experiment that could be conducted in the reactor as well.

The other theoretical case, the one calling for a constant flow, can also be simulated in a game. Increase and decrease are left up to chance and we need only see to it that the probabilities for both are the same. Even if the beads are, on the average, doubled just as often as they are removed, a fluctuation catastrophe will occur at some point and wipe out an entire population. This is impossible if increase and decrease are strictly alternated. A roulette player with limited resources who does not stop playing soon enough will experience this instability. For this reason, statisticians call this situation the "gambler's ruin." We can be sure of avoiding such ruin only if we can continually regulate quantity, but this presupposes unlimited supplies.

In the history of evolution, there were no imperatives forcing adherence to any particular regulating mechanisms. In general, we can assume that some combination of flux and extent or force regulation established itself in response to boundary conditions dictated by nature.

The situation from which evolution started was surely more a state of constant extents. Energy-rich material needed for the construction of macromolecules could accumulate relatively undisturbed. Only when "eobionts"—primordial organisms equipped with a primitive metabolism—began to multiply efficiently was this material used up. It then had to be supplied at a (more or less) constant rate from other sources. Increasing consumption transformed the "paradisiacal" conditions of nearly constant excess supplies into a situation of scarcity in which individuals lived, as it were, from hand to mouth. Plant growth is limited by the average flow of light put out by the sun in daily and annual cycles, and the ecological balance of nature has adjusted itself to this rhythm.

13.3 LIMITS

Human technology has upset this balance on many fronts. Nature's supplies of fossil fuels, which represent stores of energy from the sun, will be exhausted in the foreseeable future. Coal and oil are nonrenewable resources essential for the chemical industry. We cannot afford to burn them much longer, particularly since other fuels can be substituted for them. The economy will have to depend on recycling for supplies of many raw materials. But in order to do this, we will need greater amounts of energy.

Mankind is faced with a crucial decision. If we want to meet our future energy needs primarily by exploiting solar energy, we will have to accept a permanent state of limited flow with all the risks of instability involved. Or should we enter into what Mihailo Mesarović and Eduard Pestel[75] call a Faustian pact with nuclear energy? This step would solve the issue of supply, but would bring a number of other imponderables with it.

The sun radiates energy equivalent to 1.3 kilowatts onto every square meter of the outer layers of the earth's atmosphere. Only about one-half of that energy reaches the surface of the earth. The rest is diffused and reflected in the atmosphere or used up in maintaining atmospheric currents. If we assume ideal conditions, such as those in major desert areas, and make allowance for different times of day and different seasons, we find we have available for use on the earth's surface only about 250 watts per square meter. In the Federal Republic of Germany, where the sun shines on an average of only 30 to 40 percent of daylight hours, the wattage is considerably lower. The degree of efficiency for transforming this solar energy into a transportable form of energy should probably be estimated at 10 to 20 percent. If there is to be an average of 2 kilowatts available for every one of the Federal Republic's 50 million citizens—that is the top estimate of projected per capita requirements for 1980 and represents a total output of 100 million kilowatts—2 percent of the surface of the country would have to be covered with solar collectors. If the increased energy requirements of an economy based on recycling were taken into account, soon 7 percent of the country's territory or an area equivalent to its land surface now covered by buildings, streets, highways, etc., would be needed. The authors of the Club of Rome report suggest installing these solar collectors in the desert regions of the earth. But the transportation of this energy over great distances and across national borders would create new problems, quite apart from the fact that there are no global authorities to administer and control these global measures.

We should not hesitate to go ahead with projects of this kind wherever they are feasible. But in view of the massive dimensions involved and the many unsolved problems of solar energy, it seems to us neither appropriate nor responsible at this point to dismiss the "Faustian pact" altogether. Because the European nations are not richly blessed with natural resources, an asceticism of this kind would be equivalent to a death sentence for them as industrial nations.

There is much mention these days of the risks and dangers involved in the utilization of nuclear energy. We cannot and should not play them down. There is a danger that an accident could contaminate our environment, and there is also a risk in producing bomb material that could be misused. Such material is produced primarily in fast breeders (see p. 47), which are the only devices we have that promise economically feasible exploitation of nuclear energy. Whether we will ever have to make use of our total supply of radioactive fuel will depend entirely on whether we succeed in solving the technical problems posed by the supposedly less dangerous fusion reactor. We would then no longer have to worry about fuel supplies, but we still would not be relieved of all risks. One major environmental problem that is connected with any kind of energy production is that all energy is ultimately transformed into heat and has to be absorbed by nature. All these problems we have mentioned are serious, but they can be solved.

In fact we must also ask what risks would we run if we slowed down or even completely halted the development and testing of nuclear power plants. The key point is not whether there are risks but rather which risks are greater, those of going ahead or those of not going ahead. The next generation will have to live with the consequences of our decisions. It will then be difficult, if not almost impossible, to make up for our mistakes and omissions.

There will never be a future completely devoid of risks. It would be an illusion to believe that a world population of billions could live without risks, even if it adopted a "no-growth" economy. We cannot exclude the possibility that the earth could be struck by a huge meteor and experience a catastrophe comparable to that of atomic war. This danger—as Wolfgang Gentner has reported—is minute, but it does exist. And a major accident in a nuclear power plant, too—the experts calculate—is not likely to happen, on the average, more than once in millions of years.

Finally, we would like to say a word or two in response to the many recent studies prompted by the Club of Rome. We have in mind here two studies in particular. One is the model that Jay W. Forrester and his colleagues at M.I.T. developed and that forms the basis for the "first" growth study, published by Dennis L. and Donella H. Meadows.[73] The other is the so-called second Club of Rome report, written by Mihailo Mesarović and Eduard Pestel.[75] The authors of these reports deserve full credit for reaching a large public throughout the world and shaking people out of their lethargy. But even these authors are unable to say what should be done now that will not endanger the stability of the whole system.

It would be of great benefit if these reports could make clear to the younger generation

- how important it is to consider these questions objectively and in detail,
- how many exciting problems are waiting to be solved in this area, solved by new ideas and not merely by feeding data into a computer, and
- how unproductive it is to spout slogans in the political arena and to endanger the stability of society and of the entire human ecosystem by constantly calling for limitations without knowing how or what to limit to produce the desired results.

Reference is repeatedly made to natural regulation in ecological systems, but these references usually overlook how complex the solutions are even to relatively simple ecological situations, and how undesirable or inhuman their consequences are. If we desired no more than an ecological solution, all we would have to do is let things take their course. The resulting equilibrium, however, would not be acceptable in human terms.

Because the problem of limiting growth involves so many different phenomena, such as population increase, dissipation of energy, investment of capital, production of goods, and pollution of the environment, no undifferentiated view of this problem can do it justice.

If we listed the above items in terms of priority, population control would still head the list. It contains the key to solving all the other problems. We have to be clear about that. It is an oversimplification to blame environmental degeneration solely on increasing technology and industrialization. We should consider what environmental problems a city inhabited by millions of people would have if everyone who now drives to work in a car came on horseback instead. The only answer to growing environmental pollution lies in technology. We are using the word "technology" in its broadest sense here. It may well include biological and ecological controls guided by men.

Even if population growth were stopped completely, we would still, as a society, be faced with scarcity. At the present, human society as a whole cannot afford a halt in investment, much less a reduction in energy consumption. Such measures would be possible only if there were a drastic reduction in world population. Present trends make that a purely utopian thought. We will never see paradise again. Even those who dream of a classless society are beginning to understand this. A society of scarcity can, at best, change its class structure. It will have to remain a society oriented to achievement.

And here we come to still another problem that also affects us directly: the growth of power. Power, too, spreads according to its own laws, not, as Jacob Burckhardt expressed it, because power is "evil by nature"—it is most dangerous when its adherents are idealistically motivated—but simply because power is autocatalytic by nature. The more forces it unites in itself, the faster it spreads and the more stable it becomes. Once it has become established, it looks after its own survival, just as a bead distribution does in our game "Once and for All."

Our convictions arise from a process of mental optimization. If we did not consider our own opinions and decisions the best possible ones, we would not be able to identify with them. This implies, however, that we assign a value to our own opinions that is higher than the average value we allow all other opinions combined, and this is precisely where the danger for a democracy lies, because democracies function on an averaging of values.

The average of a series of numbers is always larger than the smallest number in that series, but it is also smaller than the largest one. A democracy does not operate on the conviction that its decisions are the best possible ones but on the insight that subjective human judgment is objectively limited. Expanding power functions on just the opposite basis. The main reason why power is autocatalytic is that its disciples are subjectively convinced of the purity of their own motives, quite apart from whether these motives are objectively good or bad.

A democratic state can allow its citizens every freedom—we trust our readers will forgive us this exaggeration—except the one that endangers its own existence. Democratic laws have to protect the individual's freedom and latitude of action, but at the same time they cannot allow the spread of organized power.

FOUR

In the Realm of Ideas

The truth is, my children, that we are players in a puppet show. More important than anything else in such a puppet show is to keep the author's idea clearly in mind.

Max Delbrück, Acceptance Speech for the Nobel Prize, 1969

14

Popper's Three Worlds

In dividing the real world into three basic categories, Karl Popper places man in the center and then distinguishes the subjective world of feelings and ideas from its material sources on the one hand and from its objective products, as represented in the body of human culture, on the other. John C. Eccles has studied the localization and manifestation of these three worlds in the human central nervous system. The processes of self-organization within these three worlds are based on universal mechanisms of selective evaluation.

When we speak of the unity of nature, we do not mean to imply that the complexity of reality can be captured in a sequence of all-embracing equations, much less reduced to a single universal formula. The universality of natural laws is balanced by the highly varied structure of the real world.

The real world is built up in layers. We should not concern ourselves so much with the unbridgeable gaps between these layers—for it could be that these gaps exist only in our minds—as with the finding of new laws and categories that grow out of the lower, material layers but that are clearly distinct from and independent of them.[76]

Describing this passage from Nicolai Hartmann as "beautiful," Konrad Lorenz cites it to suggest the unity he feels exists between the ontologist Hartmann and himself as ethologist and phylogeneticist.[77] The multiform levels of the real world and the differences in the categories used to describe it call for a comparative analysis; the uniform principles that govern the structure of the world are not always obvious. "Layering" requires self-organization and integration of the complex structures that arise through superposition. If we look at the *laws* behind the striving for order within different realms, we find more agreement between those laws than we find in the *structures* themselves and in what they accomplish.

Nicolai Hartmann postulates four "major layers of reality": the inorganic, the organic, the psychic, and the intellectual. Karl Popper's tripartite division[78] of the world (see Figure 54) places greater emphasis on the special position we human beings occupy as questioning beings and as objects of our own questioning—"The proper study of mankind is man."[22] The ego, the subjective world of the inquiring mind, stands, as it were, like a mirror between World 1, made up of the objects on which our questions focus, and World 3, made up of the answers we formulate and give material shape to. In addition to objects, including living beings and therefore human beings in their material aspect, World 1 also includes the relationships between objects. World 3 is made up of ideas, theories, and reflections to the extent that they are recorded, have the status of human documents, and thereby have an objective existence, like the objects of World 1.

The implications and limits of a theory become clear only when we attempt to find out what new relationships it elucidates. We must therefore explore how Popper's three worlds interact in human consciousness. John C. Eccles,[79] one of the great neurobiologists of our time and an enthusiastic adherent of Popper's three-world model, has made such an attempt by trying to localize the links between these three worlds (see Figure 55) in the human brain. But in this very

WORLD 1		WORLD 2		WORLD 3
Physical Objects and States		*States of Consciousness*		*Knowledge in the Objective Sense*
1. Inorganic Matter and energy of the cosmos		Subjective knowledge Experience of Perception		1. Cultural heritage recorded on material substances Philosophical
2. Biology Structure and actions of all organisms —human brain		Thought Emotions Plans and intentions Memories Dreams Creative imagination		Theological Scientific Historical Literary Artistic Technical
3. Artificially created objects Artifacts Of human creativity Of tools Of machines Of books Of works of art Of music				2. Theoretical systems Scientific Problems Critical Arguments

Figure 54. Tabular representation of the three worlds as Karl Popper defined them, including everything that exists as well as all experience (diagram taken from J. C. Eccles[79]).

process of objectification, the limitations to which all schematic divisions of this kind are subject become clear. World 2 is meaningful only as a subjective world of the self; as such it cannot be objectively represented. Or, to express this differently, once the phenomena of World 2 can be objectified they immediately belong to World 1 or to World 3. Still, it is quite legitimate to study phenomena from the realm of World 2 experimentally or even to simulate them with the help of models. Konrad Lorenz has recently made an attempt to see "through the looking glass" in this way.[77] These are his conclusions:

> The scientific study of the functional order that sustains human society and its intellectual life faces tasks of nearly infinite proportions. Human society is the most complex of all the living systems on our earth, and our

scientific investigation of it has barely scratched the surface of this complex whole. To express by how much our ignorance exceeds our knowledge would require a figure of astronomical proportions. But I believe that man as a species stands on the threshold of a new era and that the potential for an undreamed-of advance in human development already exists.

It is true that mankind is in a more dangerous situation than ever before. But science has provided our culture with the tools to escape, at least potentially, the decline to which all previous high cultures have fallen victim. This is true *for the first time* in the history of the world.

In the following chapters, we will be less interested in the content and limits of the three worlds than in the rules that determine the game of self-organization at different levels.

Living beings arise from unordered, non-organized matter. For this to happen a molecular language had to be developed so that information could be organized and transmitted. In addition, there had to be a genetic memory on the basis of which as complicated a program as the blueprint of a human being could develop, step by step. All these processes take place in World 1.

Learning follows an analogous process in the central nervous system of higher animals. Here, too, a means of communication, an "internal" language, is needed to transmit and digest the impressions recorded by the senses. These impressions take the coded form of patterns of electrical stimulation in the network of the nerve cells. The electroencephalogram is a weak, externally recorded echo of the constant and extremely complex communication between nerve cells. Memory, which is located in the network of switches, or synapses,

Figure 55. Diagram of different communication possibilities to and from the brain and within the brain. The most important lines of communication lead from the peripheral receptors to the sensory cortex and from there to the cerebral hemispheres. The output that connects the cerebral hemispheres with the muscles via the motor cortex is also illustrated. The corpus callosum is shown as a strong link between the left-hand, dominant side of the brain and the right-hand, subdominant side. The diagram also suggests the interaction between Worlds 1, 2, and 3. (Diagram taken from J. C. Eccles.[79])

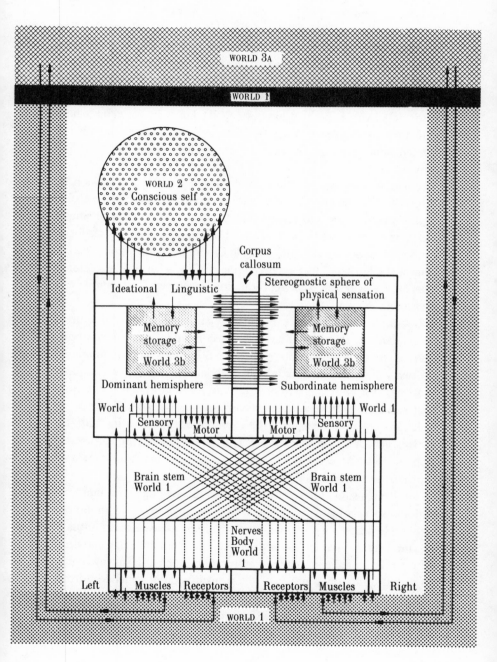

sees to it that incoming information is evaluated selectively. The engram, the constant modification of memory structure that results from this process, conditions the structure of the subjective experience of World 2.

And, finally, we come to intellectual and cultural evolution. Man is the only living being that has developed a language based on logical principles. With its help he can transmit, exchange, and recombine the limited subjective experiences and thoughts his senses have furnished him.

When we say that language and the concomitant expansion of subjective experience it provides was the decisive element in the evolution of man, we have to realize that this was not a matter of simple cause and effect but instead involved a complex chain of feedback processes. At some very early stage in evolution an analogous "extension of the horizon" had occurred once before when the recombination mechanism of sexual reproduction developed from simple cell division.

In simple cell division, evolutionary gains benefit only the direct descendants in that line of cells. The recombining mechanism, however, allows the entire species to benefit. Because of a constant mixing process whose laws Gregor Johann Mendel explained over a hundred years ago, the entire gene pool of the species is a target for mutations that can change genetic inheritance. The vast range and variety of species is the visible result of the evolutionary acceleration initiated by this genetic communication.[14] This development, too, proceeded step by step. In early phylogenetic stages we find both means of reproduction existing side by side.

The "expansion of horizons" made possible by linguistic communication displays obvious parallels. It liberates men from the limitations implicit in Darwin's principle. Because of his intensive and multifarious intellectual interactions with others, every human being can draw on the total stored experience of cultural development. In ways parallel to but quite independent of the transmission of genetic information, he can pass on this legacy from generation to generation and keep perfecting it. To be sure, this process of storing and increasing information, a process that takes place in World 3, lacks the unified, inherent evaluation that is so characteristic of the genetic learning process.

In evolution, even in its earliest subcellular stages, selective evaluation is an integral part of the contest between self-reproducing structures. The principle of evaluation inherent in the system, the principle that guarantees that the optimal structure for efficient reproduction will be selected, is causally related to the basic autocatalytic mechanism as well as to the given boundary conditions. The evaluative scale for psychic information evolved in connection with the central nervous system's assimilation of stimuli. Initially it reflected only the selection of advantageous, genetically programmed patterns of behavior. The development of evaluative centers in which pain, fear, and pleasure are localized provided greater flexibility for responding appropriately to all kinds of external stimuli. Here, too, the agreement with a selective principle oriented to survival was initially still complete. Only in man does the evaluation of psychic information definitely follow its own laws.

If we wanted to set up a similar scheme for World 3, we would have to include the many subjectively determined value scales represented by individual minds. But this very prerequisite reveals the dilemma of our social reality. There is no mechanism inherent in ideas that can automatically establish a correlation among the selective evaluations of Worlds 1, 2, and 3.

The information stored in World 3 is not automatically protected from a misuse that could result in the self-destruction of life. The survival of mankind is not guaranteed by material laws of any kind, even if the premises for that survival continue to exist. We therefore repeat: Our ethic must reflect the needs of mankind. It has to guarantee the survival of mankind without curtailing excessively the freedom of the individual. Such an ethic cannot be derived from any material laws that lie below the organizational level of man.

Intellectual dialogue with the continually changing World 3 is the task of theology. But theology, to the extent that we can call it a science, limits itself primarily to the transmission and interpretation of our religious legacy. It is more likely that contributions to morals and ethics will be made by branches of learning other than theology.

The church regards the findings of modern biology with suspicion or indifference. It has its own concept of life, and this concept has remained unchanged since the church's inception. Teilhard de Chardin[80] is probably the only modern theologian who has attempted to integrate scientific knowledge into the world view of the Christian faith. But he too has been guided more by animistic projections than by objective knowledge. There are no objectively observable signs that speak for the convergence of evolution his view propounds.

It is becoming increasingly clear in world politics that, in view of the various versions of World 2 that are the inevitable product of historically conditioned regional differences, there should be a coexistence of differing world views. But we must keep in mind here the tendency of power to expand in accordance with its own laws, a tendency we have mentioned in the preceding chapter. If we want to maintain our values, which we believe can be realized only in a free democracy, we will have to define these values in clear and unmistakable terms and see to it that they are upheld. If we fail to do this, we will always be on the defensive against any consistently applied system. Any steps, no matter how small, are irreversible if they involve a limitation of freedom.

> What everyone foresees
> Long enough
> Will happen in the end.
> Stupidity,
> A fire it is now too late to put out,
> We call fate.*

*Max Frisch, *The Firebugs.*

15

From Symbol to Language

The existence of a "language" is equally important to the material self-organization of living beings, to the communication between men, and to the evolution of ideas. The essential precondition for the development of a language is the assignment of unambiguous meanings to symbols. In molecular languages, this assigning takes the form of definite physical and chemical interactions; in communication between human beings, it is based on the assignment of meaning to phonemes and on their graphic representation. The assignment of meaning to combinations of symbols, as well as the interrelationships between such combinations, arises from an evolutionary process based on functional evaluation. According to Noam Chomsky, the deep structures of all languages—just like the genetic language that has emerged from molecular mechanisms—have common elements that reflect the functional logic inherent in the operations of the central nervous system. The parallels between molecular genetics and the generative grammar of linguistic communication make the rules affecting evolutionary processes eminently clear.

15.1 INFORMATION AND LINGUISTIC COMMUNICATION

The close relationship between the concept of in*form*ation and *form* goes deeper than etymology. Information can be understood as the abstraction of a form or gestalt, as its representation in linguistic symbols. Just as the presence of an object and its functionality are contained in the idea of a gestalt, information, too, has two complementary aspects: a quantitative aspect and a qualitative one that reveals the meaning of a given order of symbols.

This latter aspect is the more obvious in the use of language. To *inform* someone means to transmit knowledge to him. In this process, the sense and meaning of what is transmitted must be made obvious.

To comprehend the quantitative aspect—and we have already touched on this, in our discussion of entropy on pages 145-6—we have to know how much information or detailed knowledge we need to be able to identify a given order of symbols exactly. The meaning of the information contained in the symbols is not at issue here unless certain expectations related to the meaning—primarily the expectation that such a meaning exists—play a role in the evaluation of these symbols. In the simplest case, the quantitative measure of information is supplied by the number of yes-or-no decisions required to identify all the symbols in a sequence. (The best way to proceed in such a case is to put the information in question in the form of a series of binary symbols—such as 010011010111010—and then simply to count the yes-or-no decisions needed for a precise determination of the symbol sequence.) If we use binary symbols, every written symbol in the English language can be represented by a code word with five places, e.g., 00110. There are $2^5 = 32$ combinations or code words of this kind, and the code system made up by these combinations is the one the teletypewriter utilizes (see Figure 2).

If we were faced with the task of guessing a text that was completely unknown to us, we could use this method and identify every letter by asking a maximum of five yes-or-no questions. In fact, we will need far fewer than the maximum number of questions if the text to be identified is meaningful. The analysis of the game described on page 148 shows that we need no more than an average of two questions per symbol. The following reasons account for this:

- Our language uses different letters with markedly different frequency.
- There are preferred sequences of symbols (vowels, for example, will usually be framed by consonants).
- Since there is general agreement on the meaning of symbol combinations (words), many combinations can be immediately discarded as meaningless.
- Words have a certain average length.
- The laws of syntax and grammar govern the sequence and combination of words in sentences.
- Every sentence has to be meaningful (or, indeed, fulfill certain expectations).

In many cases it is possible, with the aid of these secondary clues or conditions, to decipher a sequence of symbols that contains a coded message. In such cases, entropy sinks to zero. This means that we need no additional information to identify the message. Entropy—and, consequently, the amount of information needed to identify a message —is at a maximum when every possible combination of symbols has the same a priori probability. If any secondary conditions cause a deviation from an equal distribution of symbols, the number of yes-or-no decisions needed for definitive identification will be reduced. Linguists call these conditions that reduce uncertainty *redundancies*. Certain information does in fact recur or return like waves (Latin, *unda*). In transmitting a redundant message, we could omit some symbols. Conversely, we will deliberately make a message more redundant if we fear that an overlay of "noise" will obliterate some of the information. Shannon's information theory shows how both these processes can be optimally employed.

The fact that we can distinguish at all between an absolute, quantitative aspect of information and a meaningful, semantic one can be explained by the complexity of language, and it is this complexity that still relegates the idea of a purely machine-made literature written by computers to the realm of fantasy. Competent linguists, like the Israeli mathematician and philosopher Yehoshua Bar-Hillel, are currently turning their attention more to neurobiology. Our intellectual communications system with its approximately ten billion nerve cells, each of which has in turn thousands of connections with other cells, has a nearly unlimited capacity for absorbing and combining information received from the environment. Before we can even begin to estimate its capacity, we will have to know much more about the elementary process of information storage and about its location and manifestation in the network of nerve cells and synapses. We will surely have more knowledge of this kind available to us in coming years, but even then it will not be possible to replace the creative powers of our brains with a machine.

The steps involved in linguistic communication can be represented in a diagram (see Figure 56). *A* represents a source of information with nearly unlimited capacities. Outside information picked up by the senses is evaluated and then incorporated, on the basis of programmed mechanisms, into the experience already stored in the memory. This process results in an immense capacity for assimilating information. This assimilation takes place in *B*. *C* then emits physical signals on the basis of information received from *B*. *A′*, *B′*, and *C′* represent the corresponding steps in the receiver.

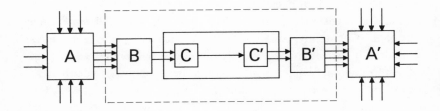

Figure 56. Communication diagram.

Communication between C and C' can include a number of technical processes, such as the coding of information for use in computers, the transmission of information over great distances, and the retrieving of information by separating out all kinds of superimposed noise. The mathematical theory of communication, often called simply information theory, finds its primary uses in this area. (Among the pioneers in this field are Ronald Aylmer Fisher, Norbert Wiener, Claude Shannon, Andrei Nikolaievich Kolmogorov, and Léon Brillouin.[42]) To the extent that information theory concerns itself with different aspects of language it also focuses on processes that take place in the centers B and B', or at least on their structural ramifications. But its applicability does not extend to the problems connected with the centers A and A', which, for the time being, seem to be the exclusive province of psychologists and philosophers.

Jürgen Habermas[81] not only distinguishes between sensory experience (observation) and communicative experience (understanding), i.e., between processes that take place in the centers B' and A', but also correctly concludes that understanding, as a kind of feedback process, has to include two steps:

> Understanding is, however, a two-step process. In the first step, it is tied to the non-objectifying aspect of the act of speech. Only if an interpersonal relation is created by performing an act of speech can we *understand* what statement or question or command, what promise, advice, etc. has been expressed. What we have understood in this non-objectifying aspect, i.e., the experience itself, is in some way objectified when we, in the next step of understanding, make it the content of an assertion.

Or, as Habermas goes on to say:

> To be able to *understand* the sentence: "Peter issues an order to John," I must at some time have learned through some experience of communication what it means to give or receive a command.

Habermas sees a paradox in "a physics the laws of which are founded on a transcendental level and which is supposed to be able to explain the transcendental accomplishments of the cognizant subject," and he bases his argument primarily on Kant's analysis of the subjective conditions needed to make experience possible.

As participants in a game we can certainly understand the rules by which it is played. To understand the rules of nature is all that physics attempts, that and no more. And there is no reason to assume that physics cannot understand the functioning of the human central nervous system in the same way. The only necessary condition is that the processes in question can be observed, and crucial insights can be gained by working with model objects. It is incorrect to think that studies of this kind are impossible because human beings are both subject and object at once here. Paradoxes usually originate in our formulations rather than in the problem itself.

Within recent years, important insights into the phenomenon of self-organization in different areas of biology have been made. The functionings of consciousness and memory in the central nervous system involve processes of this kind. Questions such as "Who does the organizing?" or "Who informs whom?"—questions that are variations on the old scholastic problem of which comes first, the chicken or the egg—prove to be just as meaningless as the search for the beginning or end of a circle. Paradoxes can usually be resolved by introducing a new dimension.

In his book *The Human Brain,* John C. Eccles has assembled current knowledge on the structure and functioning of our central nervous system.[79] According to Eccles, we have to assume that the areas depicted schematically in Figure 58 are not strictly localized and are found in both hemispheres of the brain (see Figure 57). The speech center is located in the dominant hemisphere, which is usually the left, but it is closely linked with other centers, particularly the sensory and motor centers. The corpus callosum links the two hemispheres. It is made up of about 200 million nerve fibers that are able to transmit about 4 trillion electrical impulses per second. By studying patients with damage to this organ of intercerebral communication, Roger W. Sperry was able to assign specific functions of consciousness more precisely to different regions of the brain. But we are not interested as much here in the location as in the functional manifestations of the areas pictured in Figure 55.

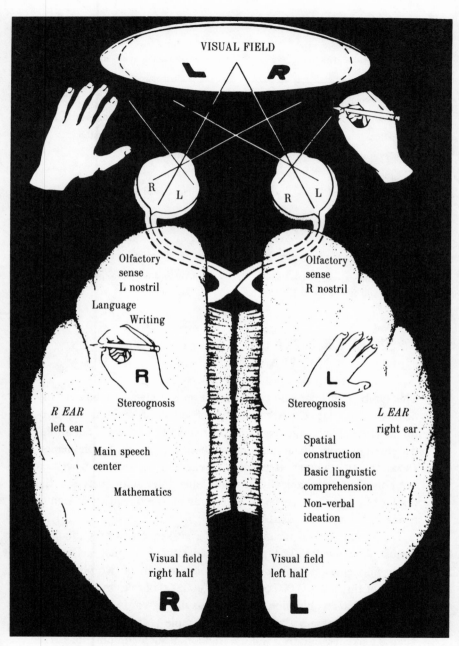

Figure 57. Stimuli from a visual field projected schematically onto the visual cortex (taken from J. C. Eccles[79]).

DOMINANT HEMISPHERE

SUBORDINATE HEMISPHERE

Link to consciousness
Verbal
Ideational
Analytical
Sequential
Arithmetical and computer-like

No such link
Almost non-verbal musical
Sense of image and pattern
Synthetic
Holistic
Geometrical and spatial

Figure 58. Localization of different abilities and functions in the two hemispheres of the brain (taken from J.C. Eccles[79]).

15.2 STRUCTURES OF LANGUAGE

The symbols of linguistic communication are always unequivocally defined. The relatively large number of letters arose from the functional needs of languages based on phonetics. The letters of our alphabet emerged from the abstraction of about fifty (or more) phonemes. In computer language, it is more efficient to use only two written symbols. The mechanically transmittable interaction between the sender (C) and the receiver (C') depends on the unequivocal significance of the symbols used.

The assignment of meaning to different sequences of symbols is almost unequivocal. The phonetic capacity of languages is huge. Forty phonemes can produce 1,600 combinations of two phonemes, 64,000 of three, 2,560,000 of four, 102,400,000 of five, and more than 4,000,000,000 combinations of six. One-syllable words can easily consist of six phonemes ("sprint," for example $[s \cdot p \cdot r \cdot i \cdot n \cdot t]$). Languages make use of only a minute fraction of such combinations. The assignment of meaning—which we described as "almost" unequivocal

—is a process that has by no means been concluded in modern languages. It remains subject to phonemic "play."

James Joyce is the acknowledged master of wordplay, but the English language in general, from Shakespeare to Shaw and on up to the present, has always had a strong tradition of puns and play. We would need many pages to list all Joyce's coinages of new words.[82] The word "quark," which Joyce made up and used in *Finnegans Wake*—[83]

> Three quarks for Muster Mark!
> Sure he hasn't got much of a bark
> And sure any he has it's all beside the mark.*

—has found a place in the exact language of science. Murray Gell-Mann adopted this word to characterize three hypothetical elementary states of matter. He hypothesized these states in an attempt to classify the plethora of previously discovered elementary particles and their symmetries.† Others of Joyce's wordplays, such as "Helterskelterpelterwelter," have great suggestive force. To "helterskelter," Joyce simply added the verbs "to pelt" and "to welter." The word is used to describe a screaming, stone-throwing mob.

The combining of words into sentences could just as well have been mentioned in connection with category B of the communication diagram in Figure 56. The diagram in Figure 59, which William G. Moulton[84] uses to illustrate the communications system of human language, shows how difficult it is to separate the realms of A and B.

The number of possible sentences we can form is astronomical. If we make up simple sentences, each consisting of one verb and two

---

*This satirical verse makes use of the saga of Tristan and Isolde. Muster (Mister) Mark is the old, senile King Mark, whom Tristan deceives. "Three quarks" is a toast—like "Three cheers"—but the word "quark" is also meant to imitate the screeching of gulls and other sea birds. To what extent Joyce—who was living in Zurich when he wrote *Finnegans Wake,* and who always liked to use phonetically interesting word combinations from other languages—had the German word *Quark* (cottage cheese) in mind is an open question.

†Gell-Mann clearly wanted to use a purely artificial word to characterize his hypothesis.

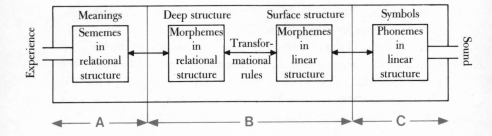

Figure 59. William G. Moulton has used the diagram shown here to characterize the communications system of human language. The similarity of this diagram to the one in Figure 56 is indicated by the blue lines.

nouns, and if we use only a thousand words from each class, we still have a total of a billion different sentences, not all of which are, of course, meaningful. Moulton calculates that if a talkative young woman kept spouting these relatively trivial sentences very fast whenever she was not sleeping or eating, even the politest observer could not call her very young by the time she had gotten through even a fraction of them. We would be dealing in "hyperastronomical" numbers if we substituted sentence structures typical of Kant. According to Wilhelm Fucks,[85] Kant is far in the lead in the use of long and complex sentences.

Noam Chomsky[86] has shown that if we disregard the specific idiosyncrasies of different languages, sentence structures display parallels that suggest universal laws apparently rooted in the organization of the human brain. Chomsky's hypotheses are supported by the preliminary work of his mentor, Roman Jacobson, and his school, as well as by findings that Eric Lenneberg[87] and others have made in observing language learning in children. These findings point to a general growth of syntax and suggest that the acquisition of language in evolution followed a similar pattern.

Chomsky bases his generative grammar on the universal applicability of a number of "internal" rules. He distinguishes between two kinds of these rules: generative and transformational rules. If we use a scheme similar to that of a decision tree (see Figure 60)—Chomsky calls this kind of diagram a phrase-marker—we can divide a sentence into a noun phrase and a verb phrase. The verb phrase can in turn be divided, for instance, into a verb and a second noun phrase (direct object). In this way, the surface structure of the sentence can be subjected to analysis or synthesis.

The "deep structure" of a complex sentence becomes clear only if we apply transformational rules. With these rules, we can separate a sentence from a surface structure that is determined by an individual language and that can be represented only by complicated phrase-markers. The sentence can then be divided into a number of simple key sentences that can be expressed by appropriate phrase-markers. For this operation we need a so-called transformation-marker or a complete scheme of individual transformations. Examples of such transformations are relative transformations (those introduced by relative pronouns), subordinating transformations, passive transformations, etc. Figure 60 illustrates a transformation of this kind.

It is debatable to what extent formalization of this kind reflects the reality of language. Critics forget too readily that while laws explain how events occur, they cannot explain specific events themselves. If we may again draw on a comparison we have used before, we could say that Chomsky's linguistics applies to language in the same way that thermodynamics does to the weather. Weather is determined by certain conditions of temperature and pressure that follow the laws of thermodynamics. Although we have long understood these laws, long-term weather predictions still depend on luck because the boundary conditions affecting weather are so complex and so difficult to ascertain.

The reality of language reflects the fact that it is still undergoing change. In his preface to Ludwig Wittgenstein's *Tractatus,* Bertrand Russell expresses some skepticism toward Wittgenstein's apodictic statement: "Whereof we cannot speak, thereof we must be silent." Russell comments, "What causes hesitation is the fact that, after all, Mr. Wittgenstein manages to say a good deal about what cannot be said, thus suggesting to the sceptical reader that possibly there may be some loophole through a hierarchy of languages or by some other exit."

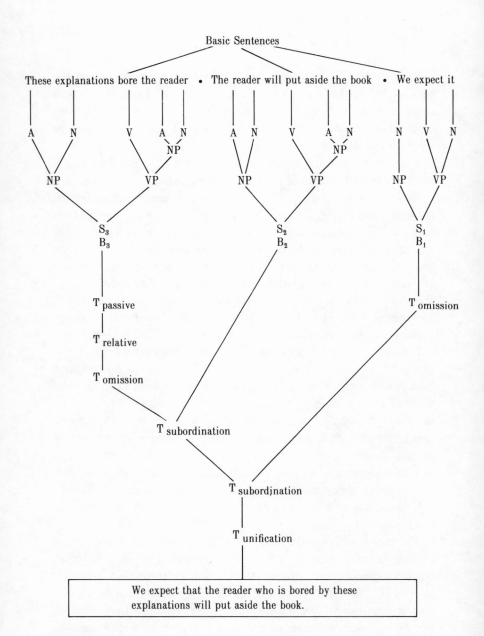

Figure 60. Chomsky's transformation of a sentence into three basic sentences. A = article, N = noun, V = verb, P = phrase-marker,

NP = noun phrase, VP = verb phrase, S = sentence, B = basic phrase-marker, T = transformation-marker.

Alfred Tarski[88] has shown the limits of using logic to assign un-equivocal meaning to language. If we want to formulate an object language whose relations are precisely defined, we need the semantics of a metalanguage. And if we wanted to objectify this language, we would need what Carl Friedrich von Weizsäcker[89] has called a meta-metalanguage, and there would be no end in sight to the hierarchy of languages that could be built up in this way.

15.3 MOLECULAR SEMANTICS

The concept of *information* has acquired central importance in molecular biology as well. Von Weizsäcker[89] notes with some surprise: "But here there is no one who speaks, no one who communicates or understands what is communicated." He then adds this qualification: "Perhaps in this case the most naive way of putting it is also the most appropriate, and this way is to use linguistic categories even where there is no speaking or hearing consciousness. Chromosomes and the growing individual relate to each other as though the chromosomes were speaking and the individual listening."

This communication between chromosomes and organism is unidirectional and amounts to an issuing of commands. Molecules speak with each other only on the phenotypical level, if at all, and here they use an "objective language" oriented to optimal functional criteria.

This phenotypical molecular language displays certain analogies to phonetically based human speech. Molecular language, too, makes use of an alphabet whose symbols are expressive in themselves, like those in language alphabets but unlike those in binary alphabets. This alphabet consists of about twenty symbols, the so-called natural amino acids, each of which has a specific chemical function. We can compare this protein alphabet to the phonemes of our language or to the letters of our alphabet abstracted from them.

The "words" of the protein language represent all the executive functions occurring in the organism: triggering of reactions, regulation, and transport. Just as in the words of our language, several—about four to eight—symbols come together in cooperative units here.

These functionally effective symbols are not just strung together linearly in the words of the protein language but are arranged in specific spatial patterns according to their chemical tasks. This is made possible by the fact that the functionally important amino acids (see Figure 11) are interspersed with foldable chain sections that have the task of precisely placing strategically important amino acids in the active center. Although this active center—the actual three-dimensional counterpart of a word in this protein language—does not include any more symbols than the verbs of our languages, a protein molecule has to contain about 100 to 500 chain elements to be able to form an active center of this kind. Each one of these molecules represents a specific activity, and we could call the enzymes the "verbs" of the molecular language because they convey what is to be done.

All the functions in an organism are precisely coordinated with each other. This means that all the words of the molecular language compose a meaningful text that can be divided up into sentences. The transmission of this text from generation to generation and the communication between the legislative and executive within the cell cannot be achieved by the proteins, whose alphabet is designed for functional efficiency.

The alphabet of the legislative language, i.e., of the nucleic acids, is constructed according to economic considerations just as the alphabet of our computers or of any mechanical transmission of information is. The alphabet of the nucleic acids uses a code word for every letter of the protein alphabet, and the association of the code word to the letter is unambiguous, but the reverse is not true. The genetic code, unlike that of the computer or teletypewriter, does not work with binary symbols but uses an alphabet that is made up of four letters and is redundant as well. This is so because the genetic code was not developed logically, as the code schemes of the machine age were, but arose naturally in correlation to the protein alphabet. For this to happen there had to be a repertoire of phenotypical characteristics much richer than the one a single complementary pair of letters could provide. A single pair of building blocks could produce only very homogeneous double-strand chains of nucleic acids. If a second pair is added, thus making two pairs of letters, the number of possible combinations in a longer sequence is greatly expanded. This allows a greater repertoire of structures of widely varying stability but does not interfere with the teleonomic impulse toward a general economizing of the transmission of genetic information and of the universal concept of the enzymatic code machinery.

The vectorial character of a chromosome's communication to an organism is regulated by a pattern that Arthur Kornberg once called the "central dogma" of molecular biology, and formulated as follows:

$$DNA \rightarrow RNA \rightarrow Protein \rightarrow Everything\ Else.$$

This is probably the most exciting short short story ever told.

DNA (desoxyribonucleic acid) is the storage device or memory for genetic information. RNA (ribonucleic acid) is the messenger that conveys the information. Protein is the executive agent that translates the information into a function and that gives rise to "everything else," i.e., the entire life process. This scheme is, of course, a gross oversimplification of what actually happens in reality, and we now know that it is not even correct in detail.

Some years ago, the Americans Howard Temin and David Baltimore found an enzyme that rewrites labile RNA messages in the form of stable DNA information. This reversing enzyme, called reverse transcriptase, was isolated from viruses that are known to cause tumors. In terms of practical applications, this discovery is of great significance, particularly for cancer research. In terms of theory, however, it came as no surprise. We have known for a long time—thanks to the work of Paul Doty and Sol Spiegelman, who also made major contributions toward the discovery of reverse transcriptase—that a single RNA chain and its complementary DNA copy can be combined into a hybrid, just as two complementary DNA strands normally form a double helix. It is likely that the molecular form of DNA did not even appear until a fairly late stage in evolution and then took over a major portion of the tasks RNA had previously performed. Because DNA has a stronger tendency to form linear double chains, it is better suited for the long-term preservation of genetic information than is its structurally less rigid and more talkative sister.

On this level, speaking, communicating, reading, and understanding involve the binding (recognizing) of the right complementary molecular building blocks (linguistic symbols) and, on the basis of information supplied, the linking together of these blocks in a macromolecular ribbon (sentence). Theoretically, this copying or transcription of information should be possible in both directions, that is, from RNA to DNA as well as from DNA to RNA, since both molecules make use of the same interactions, even though the copying machinery of the cells is so constituted that this process normally moves in only the one direction.

In chemical terms, the flow of information from RNA to protein is much more strictly established. Here there is no mechanism that simply reverses the translation; and, given the structure that proteins have, such a mechanism would hardly be conceivable. But despite this, we should not rush to a dogmatic interpretation of this principle, for there are special enzymes that are capable, without instructions from RNA templates, of manufacturing both short-chain "proteins" (or, more precisely, specific antibiotically effective polypeptids) and even long-chain RNAs with reproducible sequences.

Fritz Lipmann of Rockefeller University was able to demonstrate that certain enzymes isolated out of microorganisms can, without instructions from an RNA template and by a mechanism similar to the one Feodor Lynen found at work in the synthesis of fatty acids, string amino acids together in specific sequences and then "weld" the resulting relatively short-chained protein segments together into macromolecular rings.

Manfred Sumper of the Max Planck laboratory, without the help of an RNA template and using only an enzyme as communicating agent, has succeeded in producing long RNA chains of specific sequences as well. A few years earlier, Sol Spiegelman had isolated an enzyme complex that selectively reproduced the RNA of a bacteriophage called Qβ. The RNA of this phage obviously has a specific identifying sequence and pattern, and the copying enzyme always insists that it be shown this "passport" before it begins reproducing RNA. Sumper and Bernd Küppers were able to identify this "passport." It can be produced synthetically and attached to some other

RNA molecules, which the Qβ copying system then promptly accepts and reproduces. But this identifying part of the RNA chain hardly ever occurs in other RNAs and hence not in the natural environment of these phages, i.e., the RNA messages of its bacterial host cells. If the phage invades a coli cell, the copying enzyme manufactures almost exclusively the RNA of the phage but not any nucleic acids proper to the bacteria. The phage then reproduces itself so rapidly that the host cell deteriorates.

Manfred Sumper observed that if all RNA molecules are carefully excluded from a cell-free medium that contains, in an energy-rich form, all the building material necessary for constructing RNA, this Qβ enzyme will "knit" its own templates for reproducing RNA. It begins with short segments that are apparently produced in a sequence dictated purely by the recognition site of the protein. It then joins these segments together into long chains of several hundred links. The astounding thing about this process is, on the one hand, that the RNA chains that result from it are—for the most part, at any rate—equipped with the proper identifying characteristic and, on the other, that a clear choice is made from among the wide range of products produced. As soon as a sufficient number of different chains are present, the Qβ enzyme uses them as templates that can then reproduce these chains much more efficiently.

Christof Biebricher at Göttingen was able to prove that a large number of products arises in the primary phase, i.e., the phase of *de novo* synthesis. But a selection process sees to it that only the best-adapted sequence is chosen. We have already shown, in the context of our "Selection" game (see p. 52), how this principle works. The best-adapted sequence is the one that can reproduce itself most quickly and accurately and can achieve, at the same time, sufficient stability. All these factors affecting selective value are highly dependent on the given fold structure of the template that has the chance to reproduce. By changing the milieu, we can influence this structure. If, for example, we add to the medium an enzyme that breaks down RNA chains, then only those RNA chains whose special folds protect them against attacks by the enzyme will finally be selected. These experiments show unequivocally that limited amounts of information can also be made available by proteins. In the early periods of evolution, this very possibility, which we can simulate in an evolution reactor, must have been of great importance. The division of roles between genotype and phenotype was evidently not so clearly fixed in the early stages of life's development as it is now in the present-day products of evolution.

But still, we are not dealing here with the systematic and inherent instructional property of proteins in the sense of a reversed process of translation, i.e., a continuous reading of a protein's amino acid sequences. Genuine "bit by bit" instruction can be transmitted only in accordance with Kornberg's scheme. But let us assume that by means of a selective, *de novo* synthesis employing a primordial enzyme we could create an RNA template that would be identical with the information required for this particular enzymatic synthesizing function. And let us further assume that we had the translation machinery that could retranslate this information back into the functional protein structure. This kind of chance fulfilling of two different tasks would, of course, be quite imperfect. The probability that an optimally adapted enzyme could result in this way is so small as to be highly unlikely. The development of an effective and efficient enzyme from this imperfect *de novo* product would mean that an evolutive adaption and optimization process would have to follow immediately. But here we find ourselves in a blind alley. The optimization of enzymatic function and the selection of the RNA template that is best adapted and therefore reproductively preferred are subject to criteria that are completely independent of each other. It is true that such a *de novo* synthesis mechanism could be very important for making proteins and nucleic acids correspond, but at the same time, all the components have to be integrated into a self-reproducing system that can be selected according to unified criteria of value. A possible model for a unity of this kind is the "hypercycle" described in Chapter 12.

The gradual clarification of role division between genotype and phenotype could come about only as the translation mechanism was perfected.

The representation of phenotypical reality in genetic language—we could call this process, by analogy to the psychic function of memory, "genetic reflection"—is the result of evolution *in toto*. We are dealing here with the creation of information. In our brains, too, information can be created only evolutively, i.e., on the basis of a selective mechanism. Here, of course, the physical processes involved occur in time spans of only milliseconds.

15.4 IRREVERSIBILITY AND THE CREATION OF INFORMATION

Can information be created at all, or is it merely revealed to us? Here again we face our old dichotomy of creation and revelation. We take a distinctly different view of the evolution of life than does Jacques Monod (see pp. 166–7): We see evolution as both creation and revelation at once. Indeed, it is this very amalgamation of the two that constitutes the essence of the evolutive process.

We want to understand messages we receive. If we are to do so, a message has to reveal its meaning. It has to tie into previous experiences or agreed-upon meanings and reproduce them. But at the same time, it can also enrich our experience. Making connections, organizing, and understanding come together in an act of creation.

The distinction between the absolute and the semantic aspect of information would disappear if we could take into account and express in the probability distribution all the factors contributing to meaning. Understanding would then be the reverse process of the creation of information. It would involve a constant narrowing down of the probability distribution until finally only one alternative remained. Claude Shannon and the Hungarian mathematician Alfréd Renyi have defined the concept of informational gain quantitatively. They begin with the probability distributions before and after the reception of additional information and then calculate the average gain in information by comparing the modified individual probabilities. Identification then means that the probability of realization for all possible alternatives except one is equivalent to zero. In physics, this kind of narrowing down of a probability distribution can be achieved only through irreversible processes. A state that is initially possible and that can be characterized by a certain probability is made unstable by some unexpected event. The state breaks down, and a new situation arises in which alternatives that were previously possible are now excluded.

A selectively advantageous mutant created by a misreading of the genetic program, that is, by a statistical fluctuation, can also make a previously stable population break down irreversibly. The new information is the result of an irreversible event; it is also the result of an evaluation of meaning. Selection is, after all, made up of such evaluations. To use Karl Popper's terminology, we could say that certain previously possible alternatives are falsified. An analogous process has to take place in the brain when we read a message or make an observation.

At this point, we would like to introduce a little story inspired by Dürrenmatt's tragicomedy *The Physicists*. We are not telling this story for the sake of the plot. What interests us is a puzzle that we have incorporated into the plot. The solution to this puzzle illustrates the process of how information can be created by an unpredictable and irreversible change in a probability distribution. The story itself is, in Dürrenmatt's words, "grotesque but not absurd."

Three physicists have retreated to an insane asylum to hide their knowledge from the world and thus preclude any possible misuse of it. It is no accident that they all find themselves in this particular asylum. They have been adherents of different ideologies, and their respective governments, of whose true motives they were unaware, have sent them to the asylum to spy on each other.* Each of them has been instructed to acquire the knowledge of the two others. This would assure his government a supremacy of knowledge, which in turn would mean absolute political supremacy for that government. Each of the physicists has been forbidden, on pain of death, to reveal anything at all of his own knowledge. But all three are soon able, by indirect questions, cleverly expressed doubts, and provocative statements, to extract their colleagues' entire knowledge from them.

*In the actual play, this is true only of Eisler and Kilton, who originally have the assignment of extracting Möbius's knowledge from him.

Each of them believes, of course, that he has been the only one to succeed in this.

Although all three have succeeded in their mission, they withhold the results of their efforts from their respective governments and remain in the asylum because, in the meantime, they have all begun to doubt their governments' motives.

The three physicists are, of course, constantly shadowed by secret agents of their respective countries. If these agents should find out that a given physicist had, even unwittingly, revealed his knowledge, this would mean certain death for him. His only chance, then, would be to escape as quickly as possible. All three scientists have learned that every day at dusk there is an opportunity to flee to a neighboring neutral country.

How long can things go on this way?

One day the rumor begins to go around—and it soon reaches the three physicists—that *at least one* of them is now in possession of the others' knowledge.

At first, this rumor seems to contain no *new* information. It does not indicate anything that all three physicists have not known all along: at least one of them—and each one thinks that can only be himself—possesses the knowledge of the others. But this news does contain some new information for each of them after all: Not only does each one of them know "something," but each one also knows that the others know the same thing. This knowledge immediately sets in motion a mechanism that creates further information and automatically produces complete certainty for all three.

They each wait out one day after the rumor reaches them. Nothing happens; none of the three takes advantage of the opportunity to escape. But this is the signal for all three to make a bid for freedom as soon as possible. On the evening of the following day, they all meet in the neutral neighboring country.

How can we explain the happy ending of this story?

The rumor made an unexpected connection and triggered the following thought process in all three physicists at the same time. Assuming that one of them was still ignorant of the others' knowledge, then he would have to have fled immediately, since he was obviously the one whose knowledge the others (or at least one of them) possessed. But since *no one* fled on the first day, all of them could be sure that what had been said about at least one of them had to be true of them all, and all three of them then drew the logical conclusion.

This story is the simplest possible version of a great number of similar puzzles that usually deal with "diplomats," "cannibals," etc., instead of with "physicists."

15.5   MOLECULAR GENETICS AND GENERATIVE GRAMMAR

A comparison of the molecular and phonetic languages, such as we have undertaken in earlier sections, is meaningful only as long as the emphasis on parallels does not obscure the differences that result from their different functions.

Both languages reflect primarily the characteristic nature of the communication machinery on which they are based. In genetics, the forms statements take are sentences whose structures are determined by regulatory functions. In the operon section of a bacterial genome,

Figure 61. "Sentences" and "printer's plates" in genetic language. A typical sentence in genetic language is sketched in the right-hand side of the illustration. This kind of sentence is called an operon; it consists of the operator gene and of several structural genes related to each other. (For literature on this subject, see Hans Joachim Bogen.[90]) The reading of the entire gene unit is controlled by a repressor that does not necessarily have to be coded in the immediate vicinity of the operon it regulates. The reading produces the synthesis of an RNA message that is translated into protein language and that manifests itself in the synthesis of the enzymes coded in the gene sequence. These enzymes perform different interrelated functions in metabolism, and the metabolic products that arise from these functions are the regulating substances that either block the repressor (above right) and thereby open the way for a reading of the genes or activate the repressor (lower right) and cut off the reading. Jacques Monod and François Jacob discovered how this regulatory mechanism in the reading of genes works.

The left-hand side of the illustration shows a complete "printing plate" of a bacterial genome. The data for this drawing were compiled by W. Hayes, and the illustration itself is used with the kind permission of Carsten Bresch.[91] The abbreviations mark the positions of genes already located.

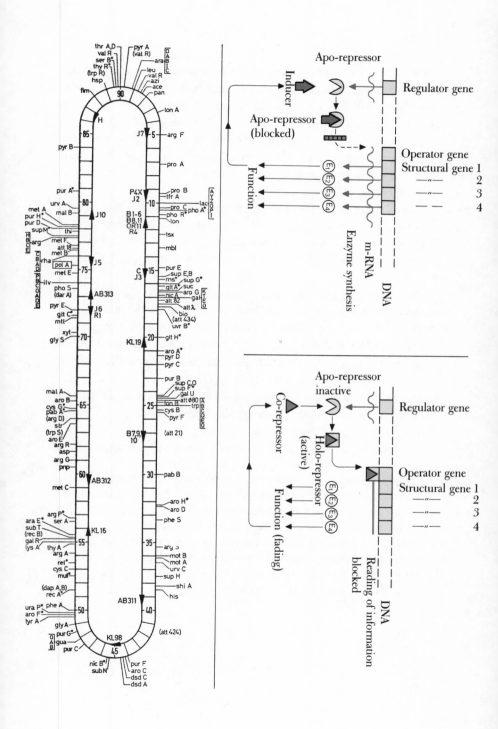

several structural genes that stand in a functional relationship to each other are joined together by a regulatory unit or so-called operator (see Figure 61). The entire genetic information of the bacterium, its genome, consists of just such sentences that are joined together in a single huge molecule. The chromosomes of more highly developed organisms have a strongly branched structure that can be seen clearly with an electron microscope, but the "syntax" of this structure is still far from being fully understood.

The sentence structure of phonetic languages also displays general structural principles. According to Chomsky, the deep structures of languages reveal a universal generative grammar that stands in a direct relationship to the generative organ of language, which is the brain.

We can say, in any case, that nature's two great evolutionary processes—the development of all forms of life and the evolution of the intellect—both depended on the existence of language. The molecular communication system of the cell is based on the reproductive and instructive capabilities of the nucleic acids as well as on the catalytic efficiency of the proteins. The resulting language, represented in the integrated functioning of this system, constitutes a new capability that cannot be directly traced back to the individual properties of its predecessors, i.e., of the isolated molecular classes.

Chomsky[86] makes a similar statement about human language:

As far as we can tell, human language results from a certain type of mental organization, not simply from a high level of intelligence. I see no reason to assume that human language is merely a more complex instance of something that can be found elsewhere in the animal world.

In this case, however, it is reasonable to assume that if we can construct empirically adequate generative grammars and determine the universal principles that control their structure and organization, then we will have made a major contribution to human psychology. . . .

16

Memory and Complex Reality

Evolution as such is comparable to a learning process based on reproductive acts of memory. When we use the word memory in connection with such varied phenomena as genetic invariance, immunity, or the psychic capability for recollection, all we are doing is making ourselves aware of the many forms that selective learning processes take. We can simulate the basic principle of such a learning process in a game.

16.1 GAMES OF EVOLUTION AND LEARNING

"The most striking characteristic of life is the vast and complex variety of forms it takes."[92]

In games, we can reduce the complexity so greatly that the optimization process will follow (nearly) predetermined paths and so result in reproducible end products. But in nature, even in the earliest phases of evolution, the number of possible combinations and, among them, the number of selectively advantageous variants was so great that we cannot trace any specific reproducible route for the historical course of evolution.

The historical route was unique. All that was established in advance was a value gradient that automatically eliminated a large number of structures at every stage of evolution. It is interesting to note here that the greater the realizable number of selectively advantageous alternatives is at any stage, i.e., the greater the probability of progress is, the less definite the individual route of evolution will be. But if there were only one realizable advantage at every stage, the mutation process that would produce that advantage would be highly improbable. Evolution would then progress at a very slow rate, but it would be completely predetermined.

If the route is uncertain, that means that the structure toward which that route proceeds is uncertain, too. There is no evidence in the results of research in molecular biology to support Teilhard de Chardin's idea of an "omega point" as a scientific thesis. The crown of the evolutionary tree consists of an unimaginably vast number of branches and twigs.

Given the enormous number of possible routes they could take, it is astonishing how rapidly evolutionary processes make their way to the goal of optimally adapted structures. We can demonstrate how this happens, in a simple case at least, with the help of what we call the RNA Game.

Table 15. T H E   R N A   G A M E

Each player has a string with a total of 80 beads on it. The red, green, blue, and yellow beads are arranged in an irregular sequence. The colors of the beads, as the name "RNA Game" suggests, correspond to the four building blocks of ribonucleic acid (RNA). The complementary colors red-green and blue-yellow represent the nucleotides that, because of specific interactions, tend to join together in complementary pairs. Two kinds of linkages—the one joining red beads to green, the other joining blue with yellow—represent the hydrogen bonds that hold the complementary bases together.

Since this is an "evolution" game, we have a mutation die in the form of a tetrahedron whose four surfaces are colored red, green, blue, and yellow respectively. In each round of play, this die is used to "mutate," in accordance with the color rolled, a bead located in a certain position.

The point of the game is, beginning with a chance arrangement of the chain, to create as quickly as possible a structure of folds that is characterized by a maximum number of complementary pairs. The roll of the mutation die applies to a bead designated by the player in advance, and only this bead can be exchanged for a bead of the color rolled whenever it seems advantageous to do so. The following rules must be strictly adhered to:

1. The Steric Rule
   Because the formation of base pairs between different sections of a sequence can be accomplished only by folding the chain on a plane, loops will inevitably result. For steric reasons, five beads in each such loop cannot undergo pair formation.
2. The Rule of Complementarity
   If two beads of complementary colors (red-green or blue-yellow) are opposite each other in the folded RNA structure and if, at the same time, rule three is fulfilled, these beads are considered a pair and are linked together.
3. The Rule of Cooperativity
   The linking of complementary beads in a pair can occur only if there is an unbroken sequence of at least four red-green pairs, two red-green pairs and one blue-yellow pair, or two blue-yellow pairs. These stable base pairs are considered "selected" and are no longer subject to the roll of the dice.

Figure 62. The RNA Game. The chain pictured in the illustration represents the chance sequence of 80 nucleotides—A (red), U (green), C (yellow), and G (blue)—that each player has at the beginning of the game. Mutations are rolled with a tetrahedral die whose surfaces are colored red, green, yellow, and blue respectively. Mutations are considered "selected" if they can form a complementary base pair (cf. rules of the game in Table 15). A comparison of the "hairpin" and "cloverleaf" structures shows that the cloverleaf pattern can derive more base pairs from the chance sequence, and therefore offers a more advantageous starting situation than does the hairpin fold.

Evaluation: The game ends after a predetermined number of rounds, or as soon as one player has a completely paired structure. The player with the highest number of points is, in either case, the winner. Every red-green pair (A = U) is worth one point; every blue-yellow pair (G≡C) is worth two. This mode of evaluation is an exact reflection of reality. The bond energy of cooperative G≡C pairs is twice as great as that of A = U pairs. But only those combinations that occur in cooperative regions count, because single pairs are unstable at room temperature.

Before the players begin to roll the die, they examine their chains for chance complementary pairs. They do this by folding the chain experimentally to see what pattern contains the most "hidden" complementary pairs. Of all patterns with both ends juxtaposed the hairpin pattern is the simplest (see Figure 62). It has only one loop and therefore offers the highest possible number of pairs. But it is important not only to form as many pairs as possible but also to construct a paired RNA structure as quickly as possible. Players soon discover that they will arrive at their goal more quickly if they use three-leaf or four-leaf clover patterns (see Figure 62) since—despite end-to-end fixation—relative leaf positions can be shifted around in many ways.

Because of the relatively high number of chance complementary pairs that these structures offer, they are most likely to provide an optimal beginning situation. Because of the rule of cooperativity, of course, each "leaf" has to contain a certain minimal number of beads, and because in each leaf five "loop" beads cannot contribute to the formation of complementary pairs, the optimal pattern of folds will depend on the total length of the chain. For chains of 80 beads, this optimal pattern is the three-leaf or four-leaf clover.

The fascinating aspect of this game is how closely it reflects a certain aspect of reality. All the rules (such as the one on the stability of complementary pairs or the one on the length of cooperative regions) are based on data established in experiments and therefore represent the behavior of RNA molecules in a thoroughly realistic way. It comes as no surprise, then, that the structures that prove victorious in this game are the very ones that have also "won" in evolution and that can be found in nature wherever the conditions of stability necessary for their selection are present. This is the way the transfer RNAs, for instance, made up of about eighty building blocks, were selected. These molecules have a purely executive function. They are responsible for the correct adaptation of an amino acid, i.e., a protein letter, to its code word in the nucleic acid language. In the early stages of evolution when the translation apparatus was formed, the nucleic acid molecules had many more functions than they do now. In that phase, the nucleic acid structures that were folded so as to produce maximal pairing were particularly resistant to being broken down and were therefore advantageous. They were better protected than single-strand molecular chains and could survive more easily. The double-strand nucleic acids that appeared in later stages of evolution and that serve exclusively for information storage were evaluated solely according to the functional effectiveness of the protein structures they represented and not according to their own structural stability.

Symmetry in the external shape of the molecule is another advantageous factor. It is easy to understand why this is so. Evolution involves many steps of reproduction, mutation, and selection. In the course of reproduction and as a result of complementary interaction, the first thing produced is a negative copy, which is then reversed into a positive in the second stage. In a symmetrical molecular structure, every stabilizing structural advantage can be utilized to the same extent by the positive and negative copies, which are related to each other like an image and mirror image. (We have already encountered a similar case of "a posteriori symmetry" in Chapter 7, p. 130.)

In our game, selection finds expression in the fact that at every roll of the dice only advantageous mutations are accepted. By rights, the chain should be presented alternately as a negative and as a positive copy. But that would be much too tedious a process. Moreover, only those structures whose ends match are admitted.

What can we learn from this game?

The fact that the winner will usually make use of a three-leaf or four-leaf clover pattern somewhat obscures the underlying complexity of the detailed structure. Even if someone played this game for a whole lifetime, no specific order of the beads would ever recur again.

If we take a hairpin structure with about forty possible complementary pairs, the probability of striking a fully paired arrangement on the first attempt is about $1:10^{24}$. For forty pairs, each of forty beads has to be matched to its complementary bead, i.e., to one out of four. There are, then, $4^{40} \approx 10^{24}$ different sequences altogether; and, conversely, for any given sequence, there is only one out of $40^{40} \approx 10^{24}$ that is exactly complementary.

In evolution, a gradient of selective value guarantees the teleonomic purposefulness that Monod also refers to. In any distribution, the one sequence that contains the maximal number of complementary pairs is the most stable and will therefore be selectively reproduced. If we momentarily disregard the restrictions imposed by the rule of cooperativity, a consistently purposeful evolution would produce a completely paired structure after an average of something less than $4 \times 40 = 160$ rolls of the dice. (We need, on the average, less than 160 steps because the random initial sequence already contains some "hidden" complementary pairs.)

We could also play this RNA Game as a game of language evolution or as a genuine game of learning. In place of forces at work between molecular building blocks, we would now have meaningful relationships between letters. Among a great number of random sequences of letters, we would always recognize and select some that could be expanded into a meaningful sentence more easily than others. Many popular games, such as Scrabble, make use of similar criteria that are rooted in the nature of our language.

Table 16. BEAD GAME "INFORMATION"

Version 1 The game uses a playing board with 64 squares (see Figure 63). Each square is occupied by a letter, a blank-space symbol, or a joker. The frequency of the letters and the blank-space symbols corresponds to the frequency of their occurrence in English. The game also calls for a pair of octahedral dice. The players roll the dice in turn, and each player places a bead of his color on the square he rolls. If this square is already occupied by one of his own beads, he may roll the dice again. But if the square is occupied by an opponent's bead that is not yet "stabilized," the player may remove this bead and replace it with one of his own. As soon as a player can form a word out of a group of letters, he can—but is not obliged to— "stabilize" the beads in question by marking them (with a sticker, for example). This means those beads can no longer be removed from the board. The player who is the first to form a complete sentence is the winner.

The rules of this game also directly reflect reality. Just as selection produces a stable structure from an arbitrary arrangement of beads in the RNA Game, it soon produces here a meaningful sentence from an arbitrary mixture of letters, a sentence whose a priori probability is extremely small. Or, to put it differently, there are certain groups of letters that are characterized by a particularly high selective value. The mutations that are produced in this version of the game by the continual discarding of "non-stabilized" beads and by the selection process that follows play a subordinate role here because the players can combine the letters as they please.

Version 2 This version makes use of a playing surface on which every group of five connected squares constitutes a code unit. Using black and white beads and following a code unknown to his opponent, one player "writes out" a sentence consisting of 20 to 25 letters (or 100 to 125 code symbols). The opponent, who is allowed to make notes, now has to figure out the sentence by guessing at the assignment of symbols. For every yes-or-no question he asks, he has to give up a bead from his supply. For every letter he guesses correctly, he receives two beads from his opponent. He may include as many letters, symbols, or words as he likes in one question. A round, of course, includes two games so that each player has a chance to guess a sentence.

The code resembles the teletypewriter code in that each letter of the alphabet is represented by five symbols. A simple recombining device can be used to vary the code. One such device consists of two paper discs that turn on a common axis. The letters of the alphabet are written on the smaller, inner disc. The code words, consisting of five symbols each, are on the larger disc. It is useful to set up the code in such a way that letters that sometimes sound alike, such as *c* and *s* or *s* and *z*, have similar code words that differ only in their last symbol. (The genetic code makes use of this principle.) By using a scheme like this, which is relatively immune to error, the players can proceed more rapidly to their goal.

It is best to begin the guessing process by identifying the symbol for the space between words. In guessing the other symbols, the players should try to make the best possible use of their knowledge about the frequency with which letters occur, word length, syntax, etc.

Version 3 The only difference between this version and the preceding one is that here the player who does the guessing already knows the code but is not allowed to see the sequence of symbols. Again, the guessing player

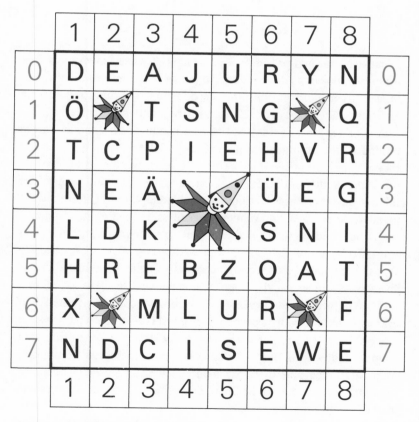

Figure 63. Playing board for a word game.

asks yes-or-no questions and has to give up one bead for each question he asks. For each symbol he guesses, or for any sequence of no more than five symbols = one letter, he receives correspondingly one or up to five beads.

This version closely resembles Shannon's linguistic game described on p. 148. A cautious player can be sure of identifying each letter with a maximum of five questions. But that will not bring him any advantage. He should therefore try to incorporate into his questions as much knowledge about the structure of the language as possible in order to minimize the number of questions he needs.

16.2   "LEARNING" NETWORKS

Because of thermal processes, every materially fixed memory is continuously subject to destruction. To preserve the information it contains, it is in need of constant reproduction. Learning requires, in addition, the capacity for teleonomically oriented modification.

A good example of this and one that is now understood at least in outline is the immune memory (see p. 79). Originally, scientists thought the immune system's ability to learn was based on a structural adaptability of the globular protein molecules or antibodies. The theory was that these antibodies could adapt to the form of the invader, the antigen, and render it harmless by binding it, precipitating it out, and breaking it down. But this theory, which was first proposed by Linus Pauling, was proved wrong when it was learned that the memory for the reproduction of proteins was located exclusively in the nucleic acids. Also, supporting evidence was soon found for a hypothesis, first presented by Frank Macfarlane Burnet, that every specific antibody is produced by a cell type of the lymph system especially programmed for that antibody. But the conclusion that seems to follow from this, i.e., that the immune memory is completely preprogrammed and therefore identical with the genetic memory, runs into a number of difficulties. The programming of the immune memory would require an enormous gene capacity. There would have to be a special antibody gene for every possible invader, that is, for every conceivable molecular structure. And even if this were possible, the remarkable "plasticity" of this memory would still not be explained.

The immune system is extremely adaptable. It does not produce any antibodies that work against the body's own structures. In addition, it can learn to tolerate foreign substances introduced either in very low or very high doses. If it were not for this learning capacity, which Peter Medawar was the first to study in detail, organ transplantations would be doomed to failure from the outset.

Thanks to Rodney Porter's studies of protein chemistry, Norbert Hilschmann's comparative work with amino acids, and Gerald Edelman's comprehensive studies of sequence, structure, and cognitive functions, the molecular structure of antibodies has been elucidated down to molecular details. Antibodies are made up of several chemically related protein chains. Every chain has so-called constant and variable regions. This means that part of the molecule is identical for all antibodies (or for certain classes of antibodies), while certain regions—especially in the vicinity of the binding center—have individual differences. Scientists assume today—and this still remains a hypothesis—that the huge repertoire of these variable regions cannot be inherited in detail because that would place too great a demand on the capacity of the genetic program. They believe instead that this repertoire is composed at a very early ontogenetic stage of the organism from a limited number of genetically programmed basic patterns. This type of somatic mutation process, for which there is some evidence, would permit the formation of an individual immune memory that tolerates the body's own protein molecules.

The different cell types of the immune system are so organized that they produce only one specific kind of antibodies and at the same time display this kind of antibody on their surfaces as receptors (see p. 79). If an antigen appears, all the cells (normally present in low concentrations) whose receptors can more or less stably bind the antigen are prompted both to produce antibodies and to multiply themselves by cell division. The receptor, then, serves as an antenna that perceives the presence of antigens, sends signals to the interior of the cell, and triggers the increased production of antibodies.

Every immune system thus has its own "language." The "vocabulary" is determined by the spectrum of cells producing antibodies. The correct use of this language is the result of a learning process and is subject to continuous modification. A memory of this kind must, of course, have some means of self-organization.

To explain this memory, Niels Jerne developed a network model that assumes that antibodies themselves have antigenic properties they exercise to produce specific reinforcing or repressing effects, such as the blocking of the "antennae." One of the main mysteries to be solved was how, given the overwhelmingly vast repertoire of molecular patterns that must include many similar and overlapping patterns, a fixed program could distinguish between the organism's own substances and foreign substances. According to Jerne, this distinguishing process is not at all necessary at first. Every protein in an organism —and therefore every antibody produced by the organism—is itself a target for another antibody molecule. A functional connection is thus established between all antibodies in the immune system, and the possibility for organized and regulated production is consequently created. These interactions link both antibody molecules and the cells equipped with receptors together into a network that responds to antigenic stimuli cooperatively and in terms of the experience it has stored. The learning capacity of the immune network results from a kind of reaction behavior that is, as we saw in Chapter 3 (see the pay-off matrix for the game of "Life and Death" on p. 28), crucial for any kind of self-organization. This behavior includes autocatalytic reinforcement and, at the same time, repression determined by population density. Processes of this kind are explicitly taken into account in the network theories of the immune system as they have been worked out in recent years by Peter Richter, Geoffrey Hoffmann, and Gerold Adam.[93]

When we hear the word "network," we usually picture an intricately interwoven structure. But in the immune network, the interweavings—like those in the molecular reproductive machinery of a cell—are of a purely functional nature.

In the central nervous system, however, the contacts between the individual nerve cells—in a human being there are more than ten

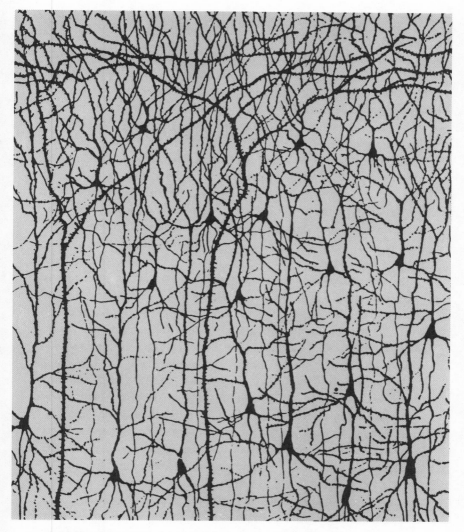

Figure 64. Working from a Golgi-stained specimen, Ramón y Cajal made this drawing of a section taken from the nerve network of a child's visual cortex. This drawing shows only a few nerve cells and their manifold contacts. This minute sample conveys some idea of the complexity of a network made up of billions of such cells.[94] The branches here differ from those in the decision tree (Figures 1 and 2) because these form innumerable interlocking cycles.

billion such cells—are made by switches, the so-called synapses. The contacts are multiple, and every neuron is linked to a large number of others (see Figure 64). These contacts transmit both chemical and electrical signals. The central nervous system is, therefore, a genuine network made up of electrical and chemical switch components. Electrical communication is, of course, much faster over great distances than chemical communication carried out by the transport of matter. Learning processes that take hours or days in the immune network are accomplished in the brain in a fraction of a second. The fixing of what has been learned or experienced is, however, a physical process. In contrast to the nearly constant and invariable number of nerve cells in the adult brain, the number of synapses is extremely variable. Synapses form and disappear and, most important, influenced by communication, keep changing the functional details of their contact, such as the threshold values for firing. The process of creating contacts between the brain cells begins in the embryo and is by no means finished at birth. This process is particularly marked in the postnatal phase and continues until sometime in the second or third year of life when all the main connections are established. Then a stationary state follows, but this does not mean that the process has come to a standstill. From this point on, the rates at which contacts are made and broken down are about equal. In old age, the rate of breakdown dominates. The properties, or stimulus thresholds, of the switch contacts change constantly. Memory and learning capacity are located in these network properties, that is, in the nature of the connections, not, as in the genome, in sequences of macromolecules. As a result, the information stored in the engram can be retrieved very quickly. By comparison, the reading and transcription of the genetic program, which involves macromolecular synthesis, is a slow process.

The stable storage of information in the brain is a relatively slow process and cannot be achieved—as mere communication can—in milliseconds. Stable storage requires plasticity of the contacts, and this plasticity is based in turn on chemical changes. Of all the body cells, the neurons have the most active metabolism; they are constantly producing ribonucleic acids and proteins. The neurons do not store learned information in these molecules but use them to build up a functional network and its modifications. Here, too, the conditions for self-organization are present: these switching components make use of circuits to stimulate and block each other autocatalytically.

Haldan Keffer Hartline[95] and Floyd Ratliff succeeded in tapping the individual fibers of the optic nerve of a king crab (*Limulus polyphemus*) and picking up the electrical intercommunication between the individual elements of the nerve network. Every individual light receptor (ommatidium) in the crab's compound eye responds to optical stimulation by sending an electrical signal to the next higher "switchboard," or ganglion. Lord Adrian and his colleagues were primarily responsible for developing this technique of tapping individual nerve fibers. Werner Reichardt[96] was able to show how alternating inhibition and reinforcement on higher levels of the network create sharp contours of the object perceived. Theoretically, the pattern of electrical stimuli could be optically reconstituted on the screen of a television set.

The analogies we perceive in the premises for self-organization in molecular, molecular-cellular, or intercellular networks, networks like those we find in the immune system as well as in the central nervous system, are reflected in the uniform basic structure of the theories used to describe this self-organization.* All the physicist can grasp are the laws governing relationships between events.

Without memory, without the continuous reproduction and evaluation or filtering of duplicated products, there could be neither an evolution of organisms nor one of ideas.

---

*David Marr, Jack Cowan, Hugh Wilson, and Christoph von der Malsburg have laid a theoretical basis for describing self-organization in the central nervous systems of mammals.[30] Their work draws on the classical electrophysiological studies that David Hubel and Torsten Wiesel made of orientation perception in the visual cortex of cats.

17

The Art of Asking the Right Question

We learn by experience. According to Karl Popper, deduction is the method of the empirical sciences. Theories can never be verified, but they can be falsified. The analysis of different mechanisms of falsification employed by nature shows that deduction has a quantitative aspect that is not taken into account when deduction is merely contrasted with induction.

A zoology student had succeeded in training cockroaches, and he proudly displayed the results of his long efforts to his professor.

He had his cockroaches fall in, and he gave them the command: "Forward, march!" The cockroaches marched forward. "Column left!" the student commanded, and all the cockroaches turned left.

The professor was about to congratulate the student on this remarkable accomplishment, but the student interrupted him. "Wait!" he said. "I still have to show you the most important thing."

The student picked up a cockroach from the last row, pulled off its legs, and put it back in its place. Once again he commanded: "Forward, march!"

The cockroaches marched as before, except, of course, for the one without legs. "Column left." Again, all the cockroaches turned on command, except for the one that lay where it had been placed.

The professor looked inquiringly at his student.

The student said proudly, "This experiment proves conclusively that cockroaches hear with their legs."

The particular point of this story (as indicated by the title of this chapter) is:

Cockroaches, which belong to one of the oldest insect orders, are, like their close relatives the grasshoppers and crickets, members of the orthoptera family. We know that in at least some of these insects, the auditory organs are indeed located in the shins of their front legs.

17.1 DEDUCTION VERSUS INDUCTION

Of what use are analogies?

They teach us to formulate better questions, perhaps even the right questions.

There is an analogy between the evolution of life and the evolution of ideas.

If we ask a biologist what the method of evolution is, he will reply: Selection.

Karl Popper's method for testing a theory critically is no less a process of "selection." His epistemology permits two alternatives: falsifiable or not falsifiable.

But not falsifiable is by no means equivalent to verifiable.

"Theories are never verifiable." Or, and we are more inclined to agree with this less apodictic formulation: "Experience can always invalidate an empirical scientific system."

Extrapolations of this kind applied to the origin of ideas—the view, for instance, "that there is no method for discovering something new that can be logically and rationally reconstructed"—seem inadmissible to us because they cannot be proved right or wrong by epistemological methods. We do not maintain the opposite. We merely fear that the obvious analogy to the selection phenomenon in biology could lead to false conclusions or, more precisely, to inadmissible analogical arguments and conclusions. Such arguments could, for instance, deny the existence of rationally founded intuition underlying the origin of ideas; and it would then follow that ideas, just like mutations, are produced independent of any goal and by some kind of generator operating on chance and that it is only the process of deductive testing that subjects these ideas to a selective evaluation.

It is true that mutations occur independent of any goal. The triggering of mutations on the one hand and the deductive testing of the phenotype on the other have to be ascribed to two completely separate chains of events. All the mutants that fail to reach the selection level set by the population die out. The population itself falsifies them. But this does not necessarily verify the existing population because it may itself fall victim to the next selectively more advantageous generation.

We can see clearly that the mode of expression here is inappropriate because the question is no longer optimally adapted to the problem. Selection has more aspects than that of the merely binary logical decision of false or not false.

The concept "fittest" can now be quantified in terms of a physically measurable parameter of value. It no longer necessarily represents— as has been claimed in the past—a tautological description of the fact of survival. Only under certain boundary conditions—such as the control of flow in the evolution reactor (see Figure 53)—will the best adapted component emerge as the sole survivor. But the environmental conditions usually prevailing in the real world can also lead to a population explosion of (almost) all mutants or to simple coexistence or even to cooperation of components of different value. The historical process of evolution has developed considerable "intuition" as a result of the value-regulated linking of individual acts of selection. The conscious expression (or even the awareness) of an idea is comparable, as a total act, to an evolutionary process involving several steps. The process taking place in the brain includes not only communication between many nerve cells, communication accomplished by stimulation and inhibition, but also an evaluative filtering on the part of the hierarchically organized network. Only then does the result, the idea, present itself to the consciousness as a unified whole that is more than the sum of its parts. This mechanism does, of course, involve both a generator operating on chance and the falsifying evaluation of the generator's signals. But the fact that the entire process is based on a hierarchical scheme of organization and not on a simple "either-or" pattern deprives the distinction between deductive and inductive of any real meaning.

Falsification can be achieved in a number of ways. Theseus used the simplest method to find the Minotaur in the labyrinth of Knossos. Ariadne gave him a thread long enough to reach to the middle of the labyrinth. She wanted to make sure that, after winning the battle, her lover would find his way back to her. All Plutarch[97] tells us is that Ariadne instructed Theseus how he could find his way through the labyrinth with the aid of the thread. The Göttingen mathematician and pedagogue Walter Lietzmann[98] wondered what these instructions consisted of. He came to the conclusion that the thread simply served to falsify the routes that Theseus had tried. Theseus had to mark each one and then roll the thread up again so that he could use it to try still other routes and finally reach his goal.

A negative selection process of this kind leaves many alternatives open for each decision. Methods based on comparative selection are much more efficient. Selective evaluation eliminates not only what has already been tested but also everything else that is not, by comparison, clearly advantageous. Selective criteria will then be used again to choose the optimal variant from among the advantageous alternatives, provided that the path to the goal is marked by some kind of value gradient. There is no reason why this process, too, cannot be called deductive. But are not the differences in degree of deduction that we observe here precisely what we mean by the concept of induction?

This is the case even more when the changes themselves are subject to goal-oriented regulation. Illustrations of this are provided by games like chess or go, in which the mutation element—an opponent's unpredictable move—can be directed by one's own initiative. The mental game of the empirical sciences surely comes under this category of deduction.

We should certainly not lend too much credence to analogies, and we could be accused of failing to distinguish clearly enough between the psychology and the logic of epistemology. But Popper's statement "In my view there is no induction"* practically forces us to point out these analogies even if they bring together such seemingly unrelated problems as the evolution of life, the genesis of ideas in the central nervous system, and the critical selection of theories in science.

Our questions must bring out the contours of our problems more clearly, just as contours are automatically sharpened in the cortical image of our sense perceptions. In physics, the sharpening of contours can be achieved by our choice of specific boundary conditions.

17.2 THE EXPERIMENT

Not every experiment can be set up to answer a yes-or-no question. Absolute questions of this kind presuppose a great deal of knowledge about the nature of the problem and can be asked only when "nature" has "accepted" us as discussion partners. The Michelson-Morley experiment, which was designed to find out if the motion of the earth influenced the speed of light, was just such an *experimentum crucis.* †
The unequivocal "no" answer that came out of this experiment cleared the way for the theory of relativity. A recent comparable experiment, which we described in Chapter 7, is the parity experiment suggested by Lee and Yang and executed by Madame Wu and her colleagues.

The point of most experiments in the natural sciences is simply to gain new insights and learn new lessons from nature. To do this, the arsenal of methods has to undergo constant expansion and refinement so that the observing, measuring, and analyzing scientist can continue to press forward into new realms of space, time, and complexity.

Galileo is regarded as the founder of the method of measurement

---

*Popper is referring to the empirical sciences here.
†This expression was coined by Francis Bacon (1561–1626).

in physics. Yet it is doubtful that he actually carried out his famous experiments on gravity in the years 1589 to 1592 by dropping stones, as legend has it, from the leaning tower of Pisa. He probably worked out the correct form of the laws of gravity later, presumably—as a recently discovered document suggests—around the middle of the year 1604.* But he did not write them down until after he was sentenced, in 1633, to confinement in his country house near Florence. It was Giovanni Battista Riccioli and Francesco Grimaldi in Bologna who conducted the actual experiments that confirmed the laws of free fall.

Quantitative experimentation is by no means limited to physics. Alchemy has developed into an exact science that could perhaps best be described as molecular architecture. This science finds its clearest expression in the syntheses created by its great masters, such as Robert Woodward, Albert Eschenmoser, and Hans Muxfeldt, to name only a few. It is not so much the molecular work of art that fascinates the chemist and in the synthesis of which he often invests years of dedicated work, as in the case of vitamin B_{12}. It is instead nature's laws that he is attempting to understand "playfully" by way of a synthesis, and in the process of his work he does in fact gain quantitative knowledge of these laws, which he can then apply at any time as he sees fit. Nature shows him the way. Natural products contain the most fascinating functional properties; this is why they were selected in the course of evolution, as Vladimir Prelog, one of the founders of the great Zurich school of organic chemistry, has suggested.

Modern biology, too, is characterized by quantitative inquiry. Here, too, the first problem was to find, among the complex range of possibilities, the optimal solutions, the solutions nature preferred. The pioneering works here are the great structural analyses done by James Watson, Francis Crick, Maurice Wilkins, and Ghobind Khorana on nucleic acids or by Frederic Sanger, Max Perutz, and John Kendrew

---

*This document contradicts the widely held view that Galileo first formulated the laws of gravity incorrectly by basing them on the proportionality between speed and distance traveled instead of on that between speed and time interval.[99]

on proteins as well as Sidney Brenner's, Seymour Benzer's, and Charles Yamofsky's applications of new insights in molecular biology to genetics.

Biology now is in a situation analogous to that of chemistry around the turn of the century. At that time, the great coming together of physics and chemistry in the atomic mechanics of Niels Bohr and Ernest Rutherford and in the quantum mechanics of the Göttingen school was imminent. The chemists were the first to discover empirically the periodic system of the elements; only later was this system explained by a unified physical theory. Similarly, young molecular biologists today are still more interested in the empirical "how" than in the epistemological and theoretical "why."

In reality, the paths scientific research follows are by no means as direct as they appear in retrospect. In conclusion, we would like to illustrate with an episode from scientific research—and this is a true story—that luck plays a major role in experimentation and that even today there is, as Karl Popper puts it, "no rationally reconstructable method for discovering something new."

Some years ago the introduction of lithium therapy opened completely new perspectives on the treatment of manic and depressive patients. The Danish psychiatrist Mogens Schou made the major pioneering effort in quantitatively working out this method, which was based on observations made by the Australian John F. J. Cade.

But it was a report from Texas that caught the attention of specialists in this field. A team of scientists had published statistics indicating that towns whose drinking water contained high levels of lithium had fewer mentally ill than the national average. These two facts were brought together in the rash claim that lithium had a beneficial effect on (unspecified) mental illnesses.

It was shown only later that lithium does have the desired effect on manic and depressive patients, but no comparable effect on other mental patients.

Then, too, lithium is effective only if it is administered in precisely defined concentrations. Too small doses, even if given over long periods of time, are completely ineffectual, and too high ones are dangerously toxic. Mogens Schou found that a specific concentration of lithium ions in the blood well above normal has to be reached and maintained. As long as this artificial lithium level is kept exactly constant, the typical symptoms of the illness disappear almost completely and the patient requires no long-term clinical treatment. But as soon as the level sinks, the often extreme symptoms return with predictable regularity.

We know practically nothing about the functional mechanism of lithium ions. It seems clear that the disease is the result of an inherited defect. On the other hand, the absolute values of effective lithium concentrations suggest a specific receptor effect. These values lie above the natural concentrations but are below intra- and extracellular concentrations of other chemically related ions, such as sodium and potassium, that usually compete with lithium ions. We are probably dealing here with a compensation for the functional loss, due to congenital defect, of a natural receptor. This loss seems more related to the specifically effective alkaline earth ions, such as calcium or magnesium, that are present in small amounts than to the alkali ions sodium and potassium that are present in high concentrations.[100] Whatever the case may be, we will someday know the answers to all these extremely interesting questions.

But what about the statistics from Texas? A study in another state where there was similar drinking water (North Carolina) showed that there was no comparable correlation between the lithium content of the drinking water and manic and depressive illness. According to quantitative clinical studies, such a correlation could hardly be expected because even the highest lithium concentrations likely to occur in drinking water lie way below levels known to be effective.

18

Playing with Beauty

Rilke called music the "language that begins where languages end."
The formal structures of a language can be analyzed. This is equally
true of the expressive media of the arts. In a work of art, the strict
concept of evolution, based on optimal functionality, is left behind. The
playful character of chance is more pronounced. The expressive force
of a work of art depends solely on the genius of the artist. But just as
evolution is the result of chance and necessity, the creation of a work
of art, too, has to conform to the strict criteria of mental evaluation.
Or, as Theodor W. Adorno puts it, "Where art is nothing but play, there
is no room left for expression."

18.1 THE USE AND LIMITS OF A THEORY OF AESTHETIC INFORMATION

The human being's ability to recombine and to reflect upon what he
perceives in terms of experiences he has already stored is, despite the

physical limitations of his brain, practically inexhaustible. The central nervous system is an "open system" that continually absorbs information, filters it according to a hierarchical system, transforms it, and incorporates it into what is already present. Information is not subject to any law of conservation. Because of the dynamic processes taking place in the brain, it is created selectively, irreversibly, and evolutively. It has the character of a gestalt as that term is used in psychology[31] (see Chapter 6).

Any information that takes the form of thought is subjective. It has three somewhat overlapping aspects that stand in direct correlation to Popper's and Eccles's triadic world view. These are:

- an absolute aspect that reflects the quantity of symbols and their redundancy
- a semantic aspect that reflects meaning and significance within a framework of generally accepted agreements and assumptions
- a subjective aspect that reflects the character, personal experiences, and personal insights of the individual

Abraham A. Moles[43] calls this last aspect the aesthetic one and defines it as follows:

> Aesthetic information cannot be translated. It does not draw on a universal repertoire but on a repertoire of knowledge that the sender and receiver have in common. Theoretically, it cannot be translated into another "language" or into another system of logical signs because this other "language" does not exist. It is closely related to the concept of personal information.

The dilemma that a genuine theory of aesthetic information encounters is the same as that encountered by a theory of semantics. The dilemma is that the creation of meaning and the assignment of meaning—like any acts of perception and learning—are processes of dynamic self-organization. An adequate theoretical approach would explicitly have to contain a time variable and could not be based solely on stationary probability distributions of symbols. At the same time, it would have to take into account the potential complexity of an open system. No theory to date has satisfactorily met these criteria.

Consequently, analyses of works of art on the basis of information theory are no more than snapshots that can, at best, provide a meaningful explanation for some redundancies. Attempts are often made to compensate for this failing by an excessive mathematization of the theory. In these attempts, Claude Shannon's entropy formula is given excessive importance. Shannon's concept yields meaningful results only when it is applied to a closed system with a definable quantity of symbols and with a defined probability distribution. The objects of aesthetic investigation cannot generally be normalized, and their true informational content is lost in the process of forming average values.

Abraham A. Moles's studies on the "sociocultural originality of musical programs" illustrate this point. We do not mean to criticize the diligently prepared tables that show the probability with which certain works of music will be performed. These tables are very instructive, particularly as a reflection of constantly changing taste. Nor are we troubled by the disregard for the conditions affecting normalization in the calculation of entropy values (cf. pp. 149–50). This is a failing that could easily be remedied. What concerns us instead is the meaninglessness of an attempt to reduce to a single numerical value a complex and subtle expression arising from a number of different causes. What point is there to a "parameter of originality" that, in one case, reflects the probability with which certain highly demanding and therefore rarely played works are performed and, in another, uses the same value to represent the combination of a popular, frequently played work with an insignificant and nearly forgotten composition? An average value of this kind does not summarize information; it destroys it.

Another magical quantity of information aesthetics is George David Birkhoff's "measure of aesthetic order."[101] We could describe this quantity as subjective redundancy because it expresses the relative change in entropy that takes place in an observer as he contemplates an object. Reflecting a possibly very small difference between two average values that both contain a certain degree of error, this quantity can make only a very vague statement. It would be more useful to calculate the average value of the differences than the difference between the average values. But to do this, we would have to determine the changes of individual probability due to observation of the object and only then average out all the values of these differences.

But here we encounter the same question as before. Why calculate averages if we can perceive detailed information and also convey it? It is this information alone that is relevant to a work of art. In physics, we use the statistical concept of entropy precisely because the detailed information (i.e., the microstates) are not accessible by means of observation and because, if it incorporates a sufficiently large number of microstates, the average value adequately describes macroscopic behavior.

All this does not mean that a work of art is, by nature, completely inaccessible to analysis by information theory. This analysis, however, should not be an end in itself but should emerge from close cooperation between the analyzing scientist and the creative artist.*

As in the case of games, an analysis of art can deal only with the rules and regularities that are in part inherent in the genre of the work of art and in part determined by the style of the period and the personality of the artist.

As an example, we will examine the frequency with which certain intervals occur in a number of musical compositions. The intervals in a piece of music are more characteristic for a composer and his epoch than the distribution of the notes themselves, which—at least up to Schönberg—was essentially determined by the key in which the piece was written. The frequency distribution of intervals between successive notes is, however—not taking the mode (major or minor) into account—largely independent of key. This distribution is determined

---

*An attempt to realize this kind of cooperation—perhaps within the framework of a "Max Planck Institute for Music"—fell victim to the criterion of "relevance" that dominates our times. The members of the so-called Hinterzarten Study Group were Pierre Boulez, Frieder Eggers, Manfred Eigen, Wolfgang Fortner, Reinhold Hammerstein, Werner Heisenberg, Aurèle Nicolet, Edith Picht-Axenfeld, Georg Picht, Paul Sacher (as chairman), Friedrich Schneider, Carl Seemann, Carl Friedrich von Weizsäcker, Otto Westphal, Carl Wurster, Konrad Zweigert, as well as the chiefly advisory members Dietrich Fischer-Dieskau, Yehudi Menuhin, and Rudolf Serkin. Boulez has achieved a part of this group's plans in the context of the Centre Beaubourg that is being formed in Paris.

more by the kind of composition and reflects primarily the role of the part in question, whether melody or accompaniment.

Figure 65 summarizes the results of some of the analyses made by Wilhelm Fucks.[85] In each case, Fucks studied the violin part. In classical works of chamber music, the violin usually carries the melody, but it does not display brilliant scales and passages to the same degree as it does in concertos. This point is important for the evaluating of results.

To represent the correlation between notes, Fucks chose a scheme familiar to us from distance charts on road maps. In such a chart, the cities in question are listed in the same order on the two coordinate axes so that—apart from the main diagonal—each position on the table is associated with two cities, and thereby states the distance between them. In the correlograms shown in Figure 65, the size of the dots represents the frequency with which different intervals occur. But in contrast to the diagonal in the distance chart, the one here conveys meaningful information: The same note can recur several times in succession; indeed, in many compositions, the prime is one of the most frequently occurring intervals. It is interesting to note that in Anton Webern this zero interval (see the main diagonal in Figure 65) is the only striking correlation, apart from the eleventh and thirteenth half-tone intervals. Only on closer observation do we notice other correlations that differ from those in a chance distribution (see Figure 65). These are again the major sevenths (11 half tones) and the minor ninths (13 half tones), now separated from the main diagonal by two octaves.

These correlograms also differ from distance charts in having different information in the two halves above and below the diagonal. A specific upward movement always occurs with different probability than its reverse, even though the distance from a to b is the same as that from b to a. Jean Philippe Rameau long ago pointed out the leading role of the descending fifths in the progression of the bass (see below).

But, essentially, the analyses shown here reveal an obvious fact. It can be summarized as follows:

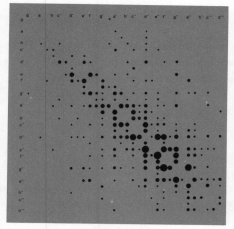

Interval frequency in Bach's
concerto for two violins (first violin)
Number of elements: 1,000
Proportion of non-zero elements: 23%

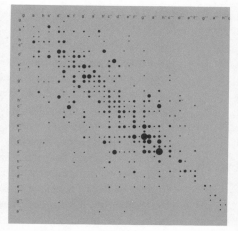

Interval frequency in Beethoven's
string quartet, Op. 74 (first violin)
Number of elements: 1,000
Proportion of non-zero elements: 16%

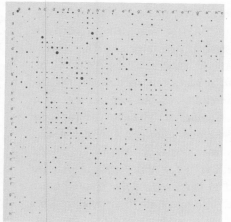

Interval frequency in Webern's
string trio, Op. 20 (violin)
Number of elements: 635
Proportion of non-zero elements: 24%

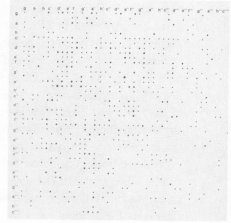

Interval frequency in
a completely random sequence of
notes
Number of elements: 635

Figure 65. Wilhelm Fucks's
correlograms for interval relations.[85]
In each case, successive notes in the
violin part were correlated two at a
time.

1. In contrast to modern music, particularly the so-called New Music, classical music shows a clear preference for certain intervals. In calling for the equal use of the twelve half tones of the octave, Arnold Schönberg has, as it were, "socialized" the intervals that were subject to a hierarchical order in classical music. But as often happens in a revolution, the succeeding order soon merely reverses the old one. As Figure 65 shows, Webern accords dissonances—especially the eleventh and thirteenth half-tone intervals—privileges over the other intervals, including the octave.

2. The uneven distribution of interval frequencies as shown in the correlograms for Bach's double concerto and for Beethoven's string quartet has to be interpreted with caution as far as its quantitative statement is concerned. The preference for the major second reflects the use of scales in the solo violin part of the concerto, but it gives us little insight into the style of the composer or the period. We cannot conclude from it that Bach, for instance, had a special preference for the second or, even less, that Beethoven favored the prime.

Lejaren A. Hiller and Calvert Bean have made an analysis of Mozart's piano sonata in C major (K 545) in which they record the absolute distribution of notes instead of the intervals. The average frequency ratio for the notes of the C major scale c : d : e : f : g : a : b is 29 : 5 : 19 : 7 : 40 : 12 : 8. (The notes f sharp, c sharp, g sharp, d sharp, and b flat occur in some modulations, but seen in the context of the ratios just listed, they are insignificant.) The above ratios show the overwhelming importance of the dominant (g). The reason why the hierarchy of the remaining notes is not as obvious is that all the notes have been considered together. We would get more detailed results if we analyzed the individual parts separately, ignored grace notes, and took certain emphases into account. If we analyzed the intervals of individual parts in the same sonata, we would find that the second is characteristic for the brilliant passages and scales of the right hand, while thirds, fourths, fifths, and sixths are found much more frequently in the sequence of bass notes in the accompaniment.

The authors of the analyses we have mentioned are, of course, fully aware of these points. Wilhelm Fucks says explicitly:

The musicologist will, of course, respond to our frequency distributions with a number of critical questions: The importance of the first violin can vary in different parts of a composition. In one part it may determine melody, rhythm, and harmony, and in another it may play a musically subordinate role. The branch of musicology concerned with quantitative research must, of course, take these and other important facts into consideration.[85]

18.2   RAMEAU AND SCHÖNBERG

What is new is not always good, and what is good is often not new. In 1722, the publishing house of Jean-Baptiste-Christophe Ballard in Paris published what probably remains the most important work on the analysis of aesthetic information to date: Jean Philippe Rameau's *Traité de l'harmonie*. [102] This book codified the laws of harmony for nearly two centuries by reducing them to their natural principles. Rameau's view that melody is based on harmony has been reduced to the absurd in the "series" of New Music, which have established new rules for relationships between tones. But the hope that the total dissolution of harmony, which we called the "socialization" of the intervals, would open the way for a greater "unity of order" in music still remains unfulfilled. This hope may yet be realized, although even now, a hundred years after Schönberg's birth, most people still perceive the kind of music he originated as "new" music. Rudolf Stephan writes in this connection:

> Perhaps Schönberg meant that the twelve-tone technique as such would make it possible to develop larger forms, but he also may not have been willing to separate twelve-tone technique clearly from thematic work. At any rate, all that Schönberg and his followers have been able to demonstrate is that twelve-tone technique combined with rigorous control of motifs does not preclude larger forms.[103]

The sensory and, above all, the psychophysical premises of our perceptual and mnemonic capabilities do not impose any natural limits on the mind's capacity to produce variations and combinations, but they do set such limits on the selective evaluation that our motivational centers demand. It is interesting to note in this connection that recent research in linguistics, too, postulates the existence of universally valid principles at work in the deep structure of languages (see p. 268). This comparison seems admissible to us, even though linguistic and musical functions—the first being primarily concerned with analysis and sequence, the second with synthesis and overall shape (gestalt)—originate in different hemispheres of the brain. As we go ahead here to emphasize certain regularities in music, we will be doing so with the full awareness that they do not constitute the complete aesthetic information conveyed by a work of art. Indeed, they are related to that information as the rules of a game are to the actual playing of that game.

In formulating his laws of harmony, Rameau focused primarily on the bass. The progression of the notes in the bass does in fact follow predictable laws of probability. Before computers were available for this kind of work, Allen Irvine McHose[104] analyzed and evaluated over five thousand tone progressions—the same kind of intervals that Fucks recorded in his correlograms—in the works of Johann Sebastian Bach (1685–1750), Georg Friedrich Handel (1685–1759), Karl Heinrich Graun (1704—1759)* and Georg Philipp Telemann (1681–1767). McHose's results are summarized in the following table. The different intervals are used in the progression of the lowest notes in the bass with the following relative frequencies:

	Prime	Second	Third	Fifth
J. S. Bach	16	21	11	52
G. F. Händel	6	29	6	59
K. H. Graun	6	18	12	64
G. P. Telemann	12	23	10	55
Average Value	10%	23%	10%	57%

In this study, octaves were not counted. This means that within an octave the perfect fourth can be seen as the inversion of the perfect fifth, i.e., as the combination of a descending fifth and an ascending

*Graun was musical director for the court of Frederick the Great from 1740 on. He was responsible for the enlargement and improvement of the Berlin Opera, and he supplied it primarily with his own works.

octave. The same holds true for the sixth and the third, on the one hand, and for the seventh and the second, on the other. In these four cases, we also have to take the major and minor intervals into account. The general inversion pattern within a key would then be:

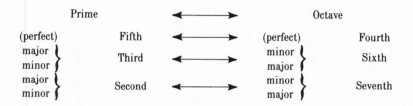

The progressions of the bass tones are vectorial, or directionally determined. Rameau's scheme deals not only with directions but also with their persistence over extended periods of time. The rules governing the choice of progression of the lowest notes are illustrated in Figure 66. McHose has extended his statistical studies to include the use of these rules by the eighteenth-century composers named above. The correlation that emerges is striking. The classifications listed in the table below are taken from Figure 66. The composers use them with the relative frequencies listed in the table:

Classification	Bach	Händel	Graun	Telemann	Average Value
1	38	42	40	35	39%
2	34	34	34	38	35%
3	19	18	18	18	18%
4	7	5	7	6	6%
5	2	1	1	3	2%

We have gone into these examples in what may be excessive detail to show not only that formal structures in musical "language" can be analyzed but also that the entire development of tonal music arose from the conscious application of the rules of harmony. But with this we have only scratched the surface of this field, and within the context of this book we can do no more than mention some other musical rules:

Classification

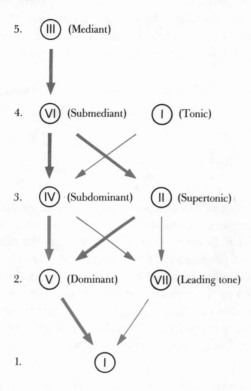

Figure 66. The progression of the bass notes is, according to Rameau's laws of harmony, largely predetermined. The illustration shows how the progression to the tonic usually occurs. From the tonic, any of the other steps can be reached directly.

The table on page 315 shows with what frequency the composers named make use of the different classifications. The normal progression shown here is used, on the average, 79 percent of the time. About 10 percent of the time, a given step is repeated before the normal progression resumes. Retrogressions, i.e., note movements reversing the normal progression, occur 7 percent of the time, the elision or omission of steps, 4 percent of the time. The thickness of the arrows indicates the frequency with which a given progression is used; the colors of the arrows indicate the intervals (red: fifth↑ or fourth; green: third↓; blue: second↑). (For additional literature on this subject, see Allen Irvine McHose.[104])

Harmony based solely on the progression of bass notes would be monotonous and boring; the possibilities for variation are too limited. This is why composers make use of modulations. Modulations can be either diatonic in nature, i.e., they can originate within the chord sequence of the key, or they can use chromatic (half-tone) motion to force transitions to more remote keys. An excessive use of modulations in connection with non-harmonic notes can ultimately destroy the tonal concept. The music of the nineteenth century, with its greater emphasis on theme and motif, underwent such a development, but without breaking completely with the principle of tonality. A similar degeneration has taken place in music "after Schönberg."

The tone sequence—particularly in the melody—of the *cantus firmus* can be made more homogeneous by introducing non-harmonic tones. These can fulfill very different functions, such as leading up to, dissolving, leading away from, supporting. Indeed, without them, there would be no such thing as melody. Ernst Bloch[105] writes:

> If we do not move with it, then no tone can move freely. It can make a few small steps on its own, but these soon come to an end. The fifth brings everything back to harmonic rest. It is only the scale that leads on, and the scale is a purely human invention.

It was counterpoint that took music beyond static harmonics. The rules of counterpoint were formulated in 1725, only three years after Rameau's *Traité* had appeared, by Johann Joseph Fux in his textbook *Gradus ad Parnassum*. A composer as recent as Paul Hindemith still refers in his music to this important work. Fux's work was translated from Latin into German in 1742, and in its importance for musical composition it is every bit as significant as Rameau's treatise on harmony.

It might seem that classical music might have exhausted the repertoire possible within the limits imposed by the theories of harmony and counterpoint. But if we look at the wealth and complexity of structures in the music of the nineteenth century, we see that its communication potential as a "language" is practically inexhaustible. We should therefore not regard New Music as a consequence of tonal music but rather as an alternative to it. This is all the more so because New Music consciously does without the element of harmony, which is physically defined and therefore directly accessible to sensory perception. Almost all the abstract elements of New Music, however, are to be understood by the intellect.

Arnold Schönberg broke completely with the harmonic laws codified by Rameau, but at the same time he set up new and equally binding rules.

Twelve-tone technique also makes use of tempered tuning, which provides for twelve half tones per octave in chromatic sequence. Because of the postulated absolute equality of all intervals (Webern, as we have seen, violated these rules; cf. Figure 65), there is no room for keys or modes. The so-called series forms the basis of New Music. In a series, all twelve tones should, if possible, occur once and only once. This would allow for about half a billion different sequences, this being the number of permutations for twelve elements: $12! = 1 \times 2 \times 3 \ldots 11 \times 12$.

This number includes all the possible combinations that can be formed by inversion, retrograde motion, and so on. By drawing on several octaves, a composer can expand his thematic range considerably. Musical ideas and inventions can thus be realized within a series —which is itself, after all, an abstract form—in a number of ways. Furthermore, in Schönberg at any rate, tone duration, rhythm, and tone color are left up to the discretion of the composer. The point of dissolving harmony is primarily to make the shaping of theme and motif absolute. This tendency assumed increasing importance in the development of classical music, too, as in Beethoven and Brahms. The dissolution of harmony no doubt means that the perception of "new" music, as compared to the "old," is moved to different centers of the brain. This suggests again that these two forms of music represent true alternatives.

Further guidelines are, of course, necessary for the thematic elaboration of series. Without such guidelines, twelve-tone technique could not, as Rudolf Stephan points out, have developed large forms. A series (or parts of a series) can appear in 48 different modifications, but there is no obligation to make use of all these possibilities. The series can be transposed onto any of the twelve half tones and take one of four basic forms: normal form, inversion, retrograde motion (also called crab motion), and retrograde inversion (inverted crab).

These variations of normal form are simple mirror images (see Figure 67). Inversion is obtained by mirroring the notes on an axis that runs horizontally through the first note of the series or parallel to that axis at a half-octave interval.

In this way, all intervals retain their absolute value, but their sign, seen in terms of the initial note, changes. In inversion, every note in the series has the same interval from the initial note as it did in the normal form, but the direction (up or down) is reversed.

Retrograde or crab motion is formed by a mirroring on the time axis, that is, on a vertical line through the last tone of the series. This mirror image consists simply of a sequence of tones read from right to left.

The crab motion of inversion and the inversion of crab motion combine the two mirroring operations, but the sequence in which the two operations are performed cannot be reversed without producing different results, assuming that the beginning and final notes of the series are not identical. (And Schönberg's rules governing series preclude such identity.) In musical notation, these two mirror axes can be readily imagined as a horizontal line running through the initial note and a vertical line running through the first or last note of the series (see Figure 67).

In New Music (after Schönberg), the application of these norms has been both extended and reduced. First of all, the idea of series was applied to other areas and thus made absolute. Metrics, rhythm, tone duration, tone color, and volume were all incorporated into serial structures. By combining different structures of these kinds, effects can be achieved that significantly influence the total sound structure. Karlheinz Stockhausen, for example, combines pitch and duration by subjecting tone duration to twelve-fold division.

György Ligeti's "field parameter" is another combination of this kind. By linking volume and interval, it produces a certain blurring of the total effect.

In general, music of the last two decades has been significantly

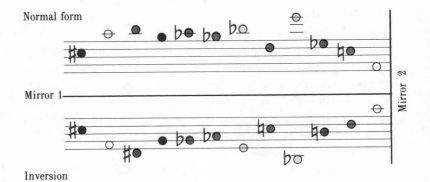

Normal form

Mirror 1

Inversion

Series from Arnold Schönberg's
Waltz (No. 5) from "Five Piano
Pieces," op. 23

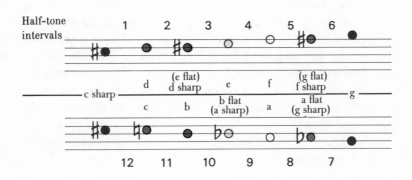

influenced by concepts from mathematics and the natural sciences. In Ligeti's music, we find the "vector" as a characteristic interval in a series. In Pierre Boulez's third piano sonata, a "formant" occurs as an optionally exchangeable section.

There are stochastic forms, developed primarily by Yannis Xenakis, that reflect the distribution laws of probability calculations. There are strategic forms based on game theory, and there are symbolic forms reflecting set theory and mathematical logic, not to mention

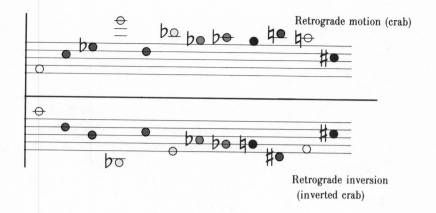

Retrograde motion (crab)

Retrograde inversion
(inverted crab)

Figure 67. The four prototypes of a Schönberg series are formed by reflections. Because of the irregular notation of half-tone intervals, the symmetries cannot be exactly reflected in the score. For this reason, we have assigned colors to the different half-tone intervals (oriented to the beginning note of the series). Like intervals have like colors, regardless of their direction. In this way, the twelve intervals of an octave can be represented by six colors. In this illustration—which could also take the form of a "glass bead game"—the four basic forms of a series are related to each other as image and mirror image. In Schönberg, the beginning tone can be transposed to any of the twelve half tones or moved to another octave altogether. If we limited the possibilities by prescribing certain reflections, new symmetry relations would result from the correlation of the beginning and final notes of a series, and this would generally have the effect of reducing atonality. The reason for this is that the inverted crab is not identical with the crab of the inversion because the first and the last notes are not identical with each other. (See Joseph Schillinger[106] for analyses of how classical music uses symmetry operations.)

John Cage's random music that draws solely on factors of chance and that results in a total dissolution of any melodic series.

No art without play! No game without rules!

Pierre Boulez, although he by no means rejects "aleatoric" music, has taken a clear position against the discarding of all rules:

> Musical composition owes it to itself to be ready with a surprise at any moment despite all the rationality we otherwise have to take upon ourselves in order to produce something worthwhile.

18.3 MUSICAL GAMES

We hope that musicians will forgive us the sketchy nature of the preceding section. We meant only to stress the importance of principles and rules in art, and to do this, we needed concrete examples.

But the essence of a work of art does not lie in the rules. Statics alone cannot produce architecture, and the fluid nature of a game demands more than the mere following of rules.

There is no lack of attempts to let the computer simulate the interplay of chance and law that goes into the creation of a work of art.

In 1956, Lejaren A. Hiller and Leonard M. Isaacson conducted a remarkable experiment at the University of Illinois. This experiment produced a suite for string quartet in four movements, the Illiac Suite, named after the computer at the University of Illinois. The experiment is described in detail in a monograph entitled *Experimental Music*. [107] In this experiment, a team consisting of a musicologist and a mathematician set themselves four tasks that they worked out in the four movements of the suite. The four tasks were:

1. with the help of the computer and by applying simplified criteria of counterpoint, to select basic themes from a random sequence of tones;
2. to combine, after the coding problem was solved, the basic themes into correct *cantus firmus* phrases by applying the rules of counterpoint successively;
3. to show that the computer is an appropriate aid to modern composition techniques because it can systematically test rhythmic structures and dynamic effects; and
4. to find and test completely new stylistic elements with the help of special computer programs.

The experiment articulated in the second movement proved to be particularly striking. From a random sequence of twelve tones, selection based on sixteen different rules of counterpoint—similar to those formulated by Johann Joseph Fux—produced step by step a perfectly organized musical movement.

Hubert Kupper,[108] using the GEASCOP program, worked out a number of similar composition games, but these had a somewhat different goal. GEASCOP means GEneral ASymptotic COmposition Program. This experiment in musical "creation" also began with a random generation of tones. Applying criteria it had "learned," the computer then selected certain tones and incorporated them into a composition. Here, the rules of composition were not fed into the computer in advance as they were in Hiller's project. The computer first had to "learn" these rules by analyzing, in terms of its own program, the tone and interval relations in selected compositions. The musical pieces programmed into the computer were divided into overlapping tone or interval sequences of defined length, and the probability distribution for the tone following on a given sequence was established and stored.

If we put aside octave relations and work with twelve half tones, we will have 1,728 three-tone sequences, almost three million six-tone sequences, and nearly ten trillion twelve-tone sequences. These figures include series with tone repetitions, a type of series that Schönberg's rules do not permit. In classical music, of course, tone repetitions are often incorporated into musical themes. The main theme of the voice part in Schubert's song "Der Tod und das Mädchen," for example, begins with sixteen repetitions of the tonic (d).

The problem in computer composition is to choose a reasonable sequence length. (Mathematicians call this length the "order of a Markov chain.") If it is too short, the computer cannot "learn" enough about the characteristic features of a composition. If the order is too long, it yields too large a number of different sequences, and the computer will produce nothing but a plagiarism in the synthesizing process.

When the computer, like an apprentice, has learned enough from its "master," it may then compose tone sequences itself. In doing this, it begins with the sequence length on which the analysis was based. This means that for every sequence of tones, the computer selects the next one to come on the basis of the "probability table" it has learned. In this way, the synthetic composition will, on the average, display— assuming it is long enough—the same tone sequences of the selected order as appeared in the model composition, but in a completely different order.

But there is one more important step, and it has to do with the word "asymptotic" in the name of the program. The application of the probability table the computer has "learned" for tone selection is regulated by a given mathematical function. In this way, certain tones or tone sequences, which were at first selected randomly, will then, depending on their frequency, be periodically favored again. This reinforcing effect somewhat resembles an abstract elaboration of theme. The sequences that were at first chosen by the computer more or less randomly are thus fixed "asymptotically" and keep returning in a pattern determined by the programmed function.

If we look at these experiments from the standpoint of the composer and of the scientist, we can make two points:

1. The computer can be of great service to the composer by giving him a chance to experiment and thus to open up new dimensions. Leonardo da Vinci would certainly not have hesitated to make use of this remarkable instrument.

2. The computer cannot realize anything but what programs devised by human beings have taught it. It is able to work through and order such material more quickly and systematically than the human mind can, but this ability is no substitute for the creativity or genius of the artist.

Composition games as they are played today by computers are by no means an invention of our times.

Bastian Perrot in all probability was a member of the Journeyers to the East. He was partial to handicrafts and had himself built several pianos and clavichords in the ancient style. Legend has it that he was adept at playing the violin in the old way, forgotten since 1800, with a high-arched bow and hand-regulated tension of the bow hairs. Given these interests, it was perhaps only natural that he should have constructed a frame, modeled on a child's abacus, a frame with several dozen wires on which could be strung glass beads of various sizes, shapes, and colors. The wires corresponded to the lines of the musical staff, the beads to the time-values of the notes, and so on. In this way he could represent with beads musical quotations or invented themes, could alter, transpose, and develop them, change them and set them in counterpoint to one another. In technical terms this was a mere plaything, but the pupils liked it; it was imitated and became fashionable in England too. For a time the game of musical exercises was played in this charmingly primitive manner. And as is so often the case, an enduring and significant institution received its name from a passing and incidental circumstance. For what later evolved out of that students' sport and Perrot's bead-strung wires bears to this day the name by which it became popularly known, the Glass Bead Game.

A bare two or three decades later the Game seems to have lost some of its popularity among students of music, but instead was taken over by mathematicians. For a long while, indeed, a characteristic feature in the Game's history was that it was constantly preferred, used, and further elaborated by whatever branch of learning happened to be experiencing a period of high development or a renaissance.

What Hesse[3] relates here in the form of a legend is historical reality. In 1793, J. J. Hummel, with imprints of Berlin and Amsterdam, published "instructions in four languages on how to compose waltzes and counter-dances with two dice." In 1806, C. Wheatstone of London came out with the identical game under the title of "Mozarts Musical Game, fitted in an elegant box, showing by an easy system [how] to compose an unlimited number of waltzes, rondos, hornpipes and reels." Whether Mozart was the originator of this game remains an open question. Some of his notebooks suggest that he was at least seriously interested in such a game (see Köchel Verzeichnis, Appendix No. 294 d). Musicologists believe that Joseph Haydn and Carl Philipp Emanuel Bach also had something to do with developing this game. In any case, a "method for shaking sonatas out of one's sleeve" was published by Johann Philipp Kirnberger in Berlin as early as 1783.

What are the principles of this game? Kirnberger's method begins by taking the bass of a known piece and finding a new melody for it. The versatility of harmony makes this a simple task. A new bass is then written for this new melody. At the same time, the key—and possibly the meter as well—is changed, and the result is a new composition that bears no obvious relation to the original one.

This is not, of course, a genuine game. But it was not long before

AN LEITUNG.

Contre-Tänze, oder Anglaises, mit 2 Würfeln zu componiren, ohne Musicalisch zu seyn, noch etwas von der Composition zu verstehen.

1) Die grosen Buchstaben A, bis H, welche über den 8. Colonnen der Zahlentafeln stehen, zeigen die 8. Tackte eines jeden Theils des Tanzes an. z. E. A, den ersten; B, den zweiten; C, den dritten; u.s.w. und die Zahlen in der Colonne darunter, zeigen die Nummer des Tackts in den Noten.

2. Die Zahlen von 2, bis 12. geben die Summe der Zahl an, welche man mit zwei Würfeln werfen kann.

3 Man wirft also z. E. für den ersten Tackt des ersten Theils des Tanzes mit 2. Würfeln 6. und sucht neben der Zahl 6. in der Colonne A, die Nummer des Tackts 105. in der Musiktafel. Diesen Tackt schreibt man aus und hat also den Anfang des Tanzes. Nun wirft man für den zweiten Tackt z. E. 8. sucht neben 8 unter B, und findet 81. in der Musiktafel. Diesen Tackt schreibt man nun zum ersten; und so fährt man fort, bis man nach 8 Würfen den ersten Theil des Tanzes fertig hat. Dann setzt man das Repetitionszeichen und geht zum zweiten Theile über.

Figure 68. The rules of the musical dice game ascribed to Mozart are very simple. They do not take into account, for instance, that if two dice are used the numbers between two and twelve appear with varying frequency. Because the middle numbers in the sequence of two to twelve will be rolled most often, we could set up the "music table" in such a way that the elements to be used most frequently are associated with these numbers. Instead of rolling the dice for set tone sequences taken from a certain composition, the players could roll for intervals that would be strung together according to Rameau's or Fux's rules. In this way, we would gradually duplicate the computer compositions described in this chapter. In this kind of composing, too, one can achieve a certain mastery, but even for this, one has to be "musical."

INSTRUCTIONS

Showing How Anyone, even if He Is Not Musical and Understands Nothing of Composition, Can Compose Counter-Dances or Anglaises with 2 Dice

1. The capital letters A to H over the 8 columns of the number table stand for the 8 measures of each part of the dance. For example, A stands for the first measure, B for the second, C for the third, and so on. The numbers in the columns indicate the number of the measure in the music supplied.
2. The numbers 2 to 12 are the possible sums that can be rolled with two dice.
3. If, for example, a player rolls a 6 for the first measure of the first part of the dance, he then looks next to the number 6 in Column A and finds the number 105, which stands for measure 105 in the music table. The player now writes out this measure and so has the beginning of the dance. He then rolls for the second measure. If he rolls an 8, for example, he will look at the number next to 8 in column B. This number is 81. The player finds this measure in the music table and copies it next to the first one. He continues in the same manner for 8 rolls, which complete the first part of the dance. Then he puts down a repeat sign and goes on to the second part.

people began to combine themes and passages from different pieces by rolling dice (see Figure 68). Rules had to be devised that made it possible for the individual pieces to be neatly fitted together. (For this purpose, Kirnberger recommends Francesco Geminiani's *Dictionaire Harmonique,* Amsterdam, 1756.) The similarity between this procedure and the composition programs described above is obvious.

And we are in fact dealing here with the same kind of bead games that we have described throughout this book. Glass beads of different colors can represent different intervals and can be arranged on strings according to Rameau's—or Schönberg's—rules.

What we then have is simply another variation of an RNA or language game that simulates the genesis of information. What kind of information is generated depends entirely on the nature of the rules underlying the game. Genetic, linguistic, and aesthetic information— no matter how much they differ in detail—all develop according to the universal principle of selective evaluation. Who is to say which rules are the more difficult? Patterns of chemical reaction and laws governing physical forces form the basis of the molecular language that has produced the genetic code. But it is a huge step from an RNA molecule to the true "works of art of evolution," insects, fish, birds, mammals, and, finally, man. And the step from Mozart's musical dice game to his Requiem is just as huge.

18.4 ART AND TRUTH
(A Dialogue That Never Took Place)

THEODOR W. ADORNO: [109]
What ultimately makes a work of art mysterious is not its composition but the truth it contains. The truth a work of art contains represents the objective solution of the mystery at the heart of each work. By demanding a solution, it points toward the truth it contains. This truth can be found only through philosophical reflection. This and only this justifies aesthetics.

SAMUEL BECKETT: [110]
If the subject of my novels could be expressed in philosophical terms, I would have had no reason to write those novels.

THEODOR W. ADORNO:
That great artists like Goethe and Beckett want nothing to do with interpretations merely stresses the difference between the truth content and the consciousness and will of the author, stresses it with the force of the artist's own sense of self. Works of art, particularly those of the highest order, call for interpretation.

THOMAS MANN: [111]
New "truth" experiences are for the artist possibilities for play and expression, nothing more. He believes in them—takes them seriously —to the extent he has to if he is to bring them to their highest expression and make the most profound possible impression with them. They are therefore a very serious matter to him, serious to the point of tears, but not entirely serious and therefore not at all. His artistic seriousness is a "playful seriousness" and absolute in nature. His intellectual seriousness is not absolute because it is seriousness in the service of play.

THEODOR W. ADORNO:
Play in art has had a disciplinary function from earliest times. It carries out the tabu imposed on expression in imitative ritual. Where art is nothing but play, expression disappears. Play is a secret accomplice of fate, a representative of the mythic burden that art would like to shake off. The repressive aspect is obvious in formulas like that of the rhythm of the blood, which is so often used to describe dance as a form of play.

JOHAN HUIZINGA: [2]
Play, we found, was so innate in poetry, and every form of poetic utterance so intimately bound up with the structure of play, that the bond between them was seen to be indissoluble. In this connection,

the words "play" and "poetry" are on the verge of losing their independent meaning. The same is true, and in even higher degree, of the bond between play and music.

THEODOR W. ADORNO:

The main criticism that can be leveled at Huizinga's thesis is that it defines art in terms of its origins. Nevertheless, his theorem is both true and false. If play is conceived as abstractly as Huizinga conceives it, then it represents nothing more specific than types of behavior that are at some remove or another from the busin of survival. What Huizinga fails to see is how much the element of play in art imitates the business of life to a much higher degree than it imitates appearances.

JOHAN HUIZINGA:

Play lies outside the reasonableness of practical life; has nothing to do with necessity or utility, duty or truth. All this is equally true of music.

CHRONICLER:

We can analyze and deduce the means art employs but not a work of art itself. We can trace the means used in a specific work, but the truth of the work depends solely on the truthfulness of the artist. Creation cannot be falsified, and the work of art mirrors nature's eternal game of creation. Art demands of the artist a total, that is to say, a playful command of the means.

List of References

1. Schiller, Friedrich von, "Über die ästhetische Erziehung des Menschen (On the Aesthetic Education of Mankind)," in a series of letters from the years 1793–94, printed in his *Philosophical and Critical Writings* (many eds.).
2. Huizinga, Johan, *Homo Ludens: A Study of the Play Element in Culture.* Boston: Beacon Press, 1955.
3. Hesse, Hermann, *The Glass Bead Game: Magister Ludi.* New York: Holt, Rinehart & Winston, 1969.
4. Fuchs, W. R., *Knaurs Buch der Denkmaschinen* (Knaur's Book of Thinking Machines). Information Theory and Cybernetics. Munich and Zurich: Droemer-Knaur Droemersche Verlagsanstalt Th. Knaur Nachfolger, 1968.
5. von Neumann, John, and Oskar Morgenstern, *Theory of Games and Economic Behavior.* Princeton, N.J.: Princeton University Press, 1963.
5a. Davis, Morton, *Game Theory: A Nontechnical Introduction.* New York: Basic Books, 1973.
6. Hassenstein, Bernhard, "Bedingungen für Lernprozesse—teleonomisch gesehen (Conditions for Learning Processes: A Teleonomic View)," Joachim Hermann Scharf, ed., in *Informatik.* Leipzig: Johann Ambrosius Barth, 1972.
7. Feynman, Richard, R. B. Leighton, and M. Sands, *The Feynman Lectures on Physics.* New York: Addison-Wesley, 1963.
8. Popper, Karl, *The Logic of Scientific Discovery.* New York: Basic Books, 1959.
9. Wittgenstein, Ludwig, *Tractatus Logico-Philosophicus.* London: Routledge and Kegan Paul, 1922.

10. Born, Max, *Albert Einstein–Max Born. Briefwechsel, 1916–1955.* Nymphenburger Verlag, Munich, 1969. Max Born, ed., *Born-Einstein Letters.* Walker & Co., 1970.

11. Macan, T. T., "Self-Controls on Population Size," *New Scientist,* vol. 28, No. 474, 1965, 801–03.

12. Darwin, Charles, *The Origin of Species,* new ed. Toronto: Crowell-Collier, 1962.

13. Spiegelmann, S., *The Neurosciences,* 2nd Study Program, ed. F. O. Schmitt. New York: Rockefeller University Press, 1970.

14. Eigen, Manfred, and Ruthild Winkler, *Ludus Vitalis,* Mannheimer Forum, 1973–74. Studienreihe Boehringer, Mannheim, 1973.

15. Sambursky, Samuel, *Das Physikalische Weltbild der Antike* (View of the Physical World in Antiquity). Zurich / Stuttgart: Artemis Verlag, 1965.

16. Landau, L. D., and E. M. Lipschitz, *Statistical Physics,* vol. 5 in the series Course of Theoretical Physics. London and Paris: Pergamon Press, 1959.

17. Perutz, M., *Röntgenanalyse des Hämoglobins* (Roentgenographic Analysis of Hemoglobin). Nobel Prizes of 1962, Imprimerie Royale. Stockholm: P. A. Norstedt and Söner, 1963.

18. Klug, A., *Assembly of Tobacco Mosaic Virus,* Fed. Proc. Vol. 31, 30, 1972.

19. Goethe, Johann Wolfgang von, *Schriften zur vergleichenden Anatomie, zur Zoologie und Physiognomik* (Writings on Comparative Anatomy, Zoology, and Physiognomy), dtv Gesamtausgabe 37. Munich: Deutscher Taschenbuch Verlag, 1962.

20. Edelmann, G., *Antibody Structure and Molecular Immunology,* Nobel Prizes of 1972, Imprimerie Royale. Stockholm: P. A. Norstedt and Söner, 1973, p. 144.

21. Gierer, A., "The Hydra as a Model for the Development of Biological Form," *Scientific American,* Dec. 1974, 44–54.

22. Goethe, Johann Wolfgang von, *Die Wahlverwandtschaften (Elective Affinities)* (numerous eds.).

23. Carter, Howard, and A. C. Mace, *Tomb of Tut-Ankh-Amen* (3 vols., reprint of 1954 ed., Cooper Square).

24. Adam, G., and M. Delbrück, "Reduction of Dimensionality in Biological Diffusion Processes," in *Structural Chemistry and Molecular Biology,* eds. N. Davidson and A. Rich. San Francisco: W. H. Freeman, 1968.

25. Schneider, D., "Kommunikation mit chemischen Signalen (Communication with Chemical Signals)," in *Max Planck Society Annual 1975.*

26. Glansdorff, P., and I. Prigogine, *Structure, Stability and Fluctuations.* London/ New York/ Sydney/ Toronto: Wiley Interscience, 1971.

27. Thom, R., *Stabilité Structurelle et Morphogénèse: Essai d'une théorie générale des modèles* (Structural Stability and Morphogenesis: An Attempt at a General Theory of Models). Reading, Mass.: W. A. Benjamin, 1972.

28. Hess, B., "Ernährung—Ein Organisationsproblem der biologischen Energieumwandlung (Nutrition—An Organizational Problem in Biological Energy Transformation)," in *Max Planck Society Annual 1974.*

29. Gerisch, G., "Periodische Signale steuern Musterbildung in Zellverbänden (The

Regulation of Pattern Formation in Cellular Complexes by Periodic Signals)," *Naturwissenschaften* 58 (1971), 430–38.

30a. Wilson, H. R., and J. D. Cowan, "A Mathematical Theory of the Functional Dynamics of Cortical and Thalamic Nervous Tissue," *Kybernetik* 13 (1973), 55–80.

30b. Malsburg, Ch. von der, "Self-Organization of Orientation Sensitive Cells in the Striate Cortex," *Kybernetik* 14 (1973), 85–100.

31. Köhler, Wolfgang, *Die Aufgabe der Gestaltpsychologie* (The Goals of Gestalt Psychology). Berlin/ New York: Walter de Gruyter, 1971.

32. Heisenberg, Werner, "Die Plancksche Entdeckung und die philosophischen Grundfragen der Atomlehre (Planck's Discovery and the Basic Philosophical Questions of Atomic Theory)," in *Schritte über Grenzen* (Steps Across Frontiers). Munich: R. Piper and Co. Verlag, 1971.

Heisenberg, Werner, "Der Begriff der kleinsten Teilchen in der Entwicklung der Naturwissenschaft (The Concept of Elementary Particles in the Development of Natural Science)," article in *Meyers Enzyklopädisches Lexicon*. Vienna/ Zurich/ Mannheim: Bibliographisches Institut, 1974.

33. Mann, Thomas, *The Magic Mountain*. New York: Alfred A. Knopf, 1956.

34. Weyl, H., *Symmetry*. Princeton, N.J.: Princeton University Press, 1952.

35. Plato, *Timaeus* (20).

36. Frauenfelder, H., and E. M. Henley, *Subatomic Physics*. Englewood Cliffs, N.J.: Prentice-Hall, 1974.

37. Dickerson, R. E., and I. Geis, *The Structure and Action of Proteins*. New York/ Evanston/ London: Harper & Row, 1969.

38. Menninger, Karl, *Ali Baba und die neununddreissig Kamele* (Ali Baba and the Thirty-nine Camels), 9th ed. Göttingen: Vandenhoeck and Ruprecht, 1964.

39. Spender, Stephen, *The Year of the Young Rebels*. New York/ Paris/ Prague/ Berlin. (R. Piper and Co. Verlag, Munich, 1969).

40. Ruch, E., "Algebraic Aspects of the Chirality Phenomenon in Chemistry," *Accounts of Chemical Research*, vol. 5, 1972.

41. Shannon, C. E., and W. Weaver, *The Mathematical Theory of Communication*. Urbana, Ill.: University of Illinois Press, 1962.

42. Brillouin, Louis, *Science and Information Theory*. New York: Academic Press, 1962.

43. Moles, A. A., *Informationstheorie der ästhetischen Wahrnehmung* (Information Theory and Aesthetic Perception). Trans. Joel E. Cohen. Urbana, Ill.: University of Illinois Press, 1966.

44. Schopenhauer, Arthur, "Parerga and Paralipomena," in *Short Philosophical Essays*, 2 vols., trans. E. F. Payne. Oxford University Press, 1974.

45. Onsager, L., *The Motions of Ions: Principles and Concepts*, Nobel Prizes for 1968, Imprimerie Royale. Stockholm: P. A. Norstedt and Söner, 1969, p. 169.

46. Meixner, J., "Die thermodynamische Theorie der Relaxationserscheinungen und

ihr Zusammenhang mit der Nachwirkungstheorie (The Thermodynamic Theory of Relaxation Phenomena and Their Relationship to the Theory of Aftereffects)," *Kolloid Zeitschrift* 134 (1953), 3.

47. Hund, F., *Grundbegriffe der Physik* (Basic Concepts of Physics). Vienna/ Zurich/ Mannheim: Bibliographisches Institut, 1969.

48. Prigogine, I., "Time, Irreversibility and Structure" (p. 561), in *The Physicist's Conception of Nature,* Lectures in Honor of Paul Dirac on his 70th Birthday, ed. J. Mehra. Boston, Mass., 1973.

49. Rhim, W. K., A. Pines, and J. S. Waugh, *Physics Reviews* B 3 (1971), 864.

50. Weizsäcker, Carl Friedrich von, "Information und Evolution (Information and Evolution)," in *Informatik,* ed. Joachim Hermann Scharf. Leipzig: Johann Ambrosius Barth, 1972.

51. Schrödinger, Erwin, *What Is Life?* Cambridge University Press, n.d.

52. Monod, Jacques, *Chance and Necessity.* New York: Alfred A. Knopf, 1972.

53. Jacob, François, *The Logic of Life: A History of Heredity.* Trans. Betty E. Spillman. New York: Random House, 1976.

54. Crick, Francis, *Of Molecules and Men.* Seattle: University of Washington Press, 1966.

55. Campbell, H. J., *The Pleasure Areas: A New Theory of Behavior.* New York: Delacorte Press, 1973.

56. Sartre, Jean Paul, *Drei Essays* (Three Essays). Frankfurt am Main/ Berlin/ Vienna: Verlag Ullstein, 1973.

57. Monod, Jacques, "L'évolution microscopique (Microscopic Evolution)," report on a lecture, *Neue Zürcher Zeitung,* February 19, 1975.

58. Bollnow, O. F., *Existenzphilosophie und Pädagogik (Existential Philosophy and Pedagogy).* Stuttgart: W. Kohlhammer Verlag, 1959.

59. Eigen, Manfred, "Self-Organization of Matter and the Evolution of Biological Macromolecules," *Naturwissenschaften* 58 (1971), 465–522.

60. Raphael, M., *Theorie des geistigen Schaffens auf marxistischer Grundlage* (Theory of Intellectual Creativity Based on Marxian Principles). Frankfurt am Main: S. Fischer Verlag, 1974.

61. Dürrenmatt, Friedrich, *The Physicists.* Trans. James Kirkup. New York: Grove Press, 1965.

62. Gurdon, J. B., and J. Bertrand, "Transplanted Nuclei and Cell Differentiation," *Scientific American* (June 1968), 24–35.

63. Stent, Gunter, "The Dilemma of Science and Morals," *Genetics* 78 (1974), 41–51.

64. Cohen, S. N., "The Manipulation of Genes," *Scientific American* (July 1975).

65. Berg, P., D. Baltimore, H. W. Boyer, S. N. Cohen, R. W. Davis, D. S. Hogness, D. Nathans, R. Roblin, J. D. Watson, S. Weissmann, and N. D. Zinder, "Potential Biohazards of Recombinant DNA Molecules," *Science* 185 (1974), 332.

66. "Experimental Manipulation of the Genetic Composition of Microorganisms," a report by a House of Commons committee under the direction of Lord Ashby (HMSO Cmnd 5880), 1975.

67. Gardner, Martin, "Mathematical Games," *Scientific American* (October 1970 and February 1971).

68. Hirsch, E. C., *Das Ende aller Gottesbeweise* (The End to Proofs of God's Existence). Hamburg: Stundenbücher, Furche Verlag, 1975.

69. Picht, G., *Wahrheit, Vernunft, Verantwortung* (Truth, Reason, Responsibility). Philosophische Studien. Stuttgart: Klett Verlag, 1969.

70. Rechenberg, I., *Evolutionsstrategie* (Evolutionary Strategy). Stuttgart-Bad Cannstatt: Friedrich Frommann Verlag (Günther Holzboog KG), 1973.

71. Demeny, P., "The Populations of the Underdeveloped Countries," *Scientific American* (Sept. 1974), 149.

72. Westoff, C. F., "The Population of Developed Countries," *Scientific American* (Sept. 1974), 109.

73. Meadows, D. L., and D. H. Meadows, eds., *Toward Global Equilibrium: Collected Papers.* New York: Wright-Allen, 1973.

74. Samuelson, Paul A., *Maximum Principles in Analytical Economics.* Nobel Prizes for 1970, Imprimerie Royale. Stockholm: P. A. Norstedt and Söner, 1971.

75. Mesarović, M., and E. Pestel, *Mankind at the Turning Point: The Second Report to the Club of Rome.* New York: E. P. Dutton, 1974.

76. Hartmann, N., *Der Aufbau der realen Welt* (The Structure of the Real World). Berlin/ New York: Walter de Gruyter, 1964.

77. Lorenz, Konrad, *Die Rückseite des Spiegels* (Behind the Looking Glass). Munich: R. Piper and Co. Verlag, 1973.

78. Popper, Karl, *Objective Knowledge: An Evolutionary Approach.* New York: Oxford University Press, 1972.

79. Eccles, J. C., *Das Gehirn des Menschen* (The Human Brain). Munich: R. Piper and Co. Verlag, 1975.

80. Teilhard de Chardin, Pierre, *La Place de l'homme dans la nature* (Man in the Cosmos).

81. Habermas, Jurgen, *Knowledge and Human Interests.* Trans. Jeremy J. Shapiro. Boston: Beacon Press, 1971.

82. Kreutzer, E., *Sprache und Spiel im Ulysses von James Joyce* (Language and Word Play in James Joyce's *Ulysses*). Bonn: H. Bouvier und Co. Verlag, 1969.

83. Joyce, James, *Finnegans Wake.* New York: Viking Press, 1939.

84. Moulton, W. G., "The Nature of Language," in "Language as a Human Problem," *Daedalus: Journal of the American Academy of Arts and Sciences* (Summer 1973).

85. Fucks, Wilhelm, *Nach allen Regeln der Kunst* (According to All the Rules of Art). Stuttgart: Deutsche Verlagsanstalt, 1968.

86. Chomsky, Noam, *Language and Mind.* Cambridge, Mass.: MIT Press, 1965.

87. Lenneberg, E., *Biological Foundations of Language.* New York: John Wiley and Sons, 1967.

88. Tarski, A., *Introduction to Logic and to the Methodology of Deductive Science.* New York: Oxford University Press, 1941.

89. Weizsäcker, Carl Friedrich von, *Die Einheit der Natur* (The Unity of Nature). Munich: Carl Hanser Verlag, 1971.

90. Bogen, H. J., *Knaurs Buch der modernen Biologie* (Knaur's Book of Modern Biology). Munich/ Zurich: Droemersche Verlagsanstalt Th. Knaur Nachfolger, 1967.

91. Bresch, C., and R. Hausmann, *Klassische und molekulare Genetik* (Classical and Molecular Genetics). 2nd edition, enlarged. Berlin/ Heidelberg/ New York: Springer Verlag, 1970.

92. Eigen, Manfred, "Leben (Life)," article in *Meyers Enzyklopädisches Lexicon.* Vienna/ Zurich/ Mannheim: Bibliographisches Institut, 1975.

93. Jerne, N. K., "The Immune System: A Web of V-domains," The Harvey Lectures, Series 70. New York: Academic Press, 1975.

94. Bodian, D., *The Neurosciences: A Study Program,* eds. G. C. Quarton, Thomas Melnechuck, F. O. Schmitt. New York: Rockefeller University Press, 1967.

95. Hartline, H. K., *Visual Receptors and Retinal Interaction,* Nobel Prizes for 1967, Imprimerie Royale. Stockholm: P. A. Norstedt and Söner, 1969.

96. Reichardt, W., "Nervous Processing of Sensory Information," in *Theoretical and Mathematical Biology,* eds. Thomas Waterman and H. J. Morowitz. New York: Blaisdell, 1965, 344–70.

97. Plutarch, "Theseus and Romulus," in *Lives of the Noble Greeks and Romans,* 3 vols. New York: E. P. Dutton, 1957.

98. Lietzmann, W., *Lustiges und Merkwürdiges von Zahlen und Formen* (Amusing and Remarkable Items about Numbers and Forms). Göttingen: Vandenhoeck und Ruprecht, 1950.

99. Drake, S., "Galileo's Discovery of the Law of Free Fall," *Scientific American* (May 1973), 84–94.

100. *The Neurobiology of Lithium,* report on a meeting of the Neurosciences Research Program, N. R. P. *Bulletin,* 1975.

101. Birkhoff, G., "A Mathematical Approach to Aesthetics," *Scientia* (Sept. 1931), 133–46.

102. Rameau, Jean Philippe, *Traité de l'harmonie (Treatise on Harmony)* (Jean-Baptiste-Christophe Ballard, Paris, 1722). English translation by Philip Gossett, New York: Dover, 1971.

103. Stephan, R., *Neue Musik* (New Music). Göttingen: Vandenhoeck und Ruprecht, 1958.

104. McHose, A. I., *Basic Principles of the Technique of 18th and 19th Century Composition.* New York: Appleton-Century-Crofts, 1951.

105. Bloch, E., *Geist der Utopie* (1923 version) (The Spirit of Utopia). Frankfurt am Main: Suhrkamp Verlag, 1964.

106. Schillinger, J., *The Mathematical Basis of the Arts.* New York: Philosophical Library, 1948.

107. Hiller, L. A., Jr., and L. M. Isaacson, *Experimental Music* (composition with an electronic computer). New York/ Toronto/ London: McGraw-Hill, 1959.

108. Kupper, H., "GEASCOP—ein Kompositionsprogramm (GEASCOP—A Composition Program)," in *Informatik,* ed. Joachim Hermann Scharf. Leipzig: Johann Ambrosius Barth, 1972, p. 629.

109. Adorno, Theodor W., *Ästhetische Theorie* (Aesthetic Theory). Suhrkamp Taschenbuch Wissenschaft 2. Frankfurt am Main: Suhrkamp Verlag, 1970.

110. D'Aubarède, G., "En attendant Beckett (Waiting for Beckett)," *Nouvelles littéraires,* Feb. 16, 1961.

111. Heftrich, E., "Zauberbergmusik (Magic Mountain Music)" (on Thomas Mann), in *Das Abendland,* Series 7. Frankfurt am Main: Vittorio Klostermann, 1975.

INDEX

Adam, Gerold, 88, 294
Adorno, Theodor W., xii, 306, 328–30
Adrian, Lord, 297
aesthetic information, theory of, 306–21
algorithms, 188
"allosteric" reaction control, 127–8
alpha helices, 121
Ambler, Eric, 118–19
analogy, 59, 299–302
animism, 166–7, 171, 258
antibodies, 79, 292–4
antiparticles, 114–16
Arber, Werner, 181
Association of German Students for
    Socialism, 133
atoms: carbon, asymmetry in, 121–3;
    unpredictability at level of, 19–21
autocatalysis, 200, 202–4;
    morphogenesis and, 90–1, 93–101; in
    reproduction, 45, 57, 64
automata, intelligent, 187–98

Bach, Carl Philipp Emanuel, 325
Bach, Johann Sebastian, 311, 312,
    314–15
backgammon, 84–7
Bacon, Francis, 302 n.
balance, 43; see also equilibrium
Baltimore, David, 273
Bar-Hillel, Yehoshua, 262
bead games, 30–65; chemical reactions
    and, 81, 85–90; cooperative
    transformation and, 71–2; Hesse and,
    4–5, 324–5; prototype of, 31; see also
    specific games
Bean, Calvert, 312
Beckett, Samuel, 18, 329
Beethoven, Ludwig van, 311, 312, 318
Bentley, W. A., 104
Benzer, Seymour, 303–4
Berg, Paul, 186
Birkhoff, George David, 308
birthrate, see population size; strategy
Bloch, Ernst, 317

Bohr, Niels, 171–2, 304
Bollnow, Otto Friedrich, 167
Boltzmann, Ludwig, 21, 143, 149, 152, 164, 170
Boltzmann constant, 146–7
Boulez, Pierre, 309 *n.*, 320, 321
boundary conditions, 22–3, 58, 151
Bouton, Charles L., 10
Boyer, Herbert W., 181
Bragg, William Henry, 127
Bragg, William Lawrence, 127
Brahms, Johannes, 318
brain, the, 188; communication and, 254–5, 264–5; selective storing of information by, 21; *see also* memory
Brenner, Sidney, 163, 303–4
Bresch, Carsten, 280–1
bridge (card game), 17
Brillouin, Leon, 158, 159, 263
Burckhardt, Jacob, 248
Burnet, Frank Macfarlane, 292

Cade, John F. J., 304
Cage, John, 320–1
Campbell, Herbert James, 165
carbon atoms, 121–3
card games, 14, 16–17
Catal Huyuk, Anatolia, 105, 106
catastrophe, 44–7; in "Survival," 65; *see also* hyperbolic growth
catastrophe theory, 23, 96
cells, morphogenesis and, 78–83
Changeux, Jean Pierre, 127
checkers, 16, 17
chemical reactions, games based on, 81–90
chess, 16, 17, 236
children's play, 17–18
chirality, 121–4, 139
Chomsky, Noam, 259, 268–70, 282

Clausius, Rudolf, 141–2, 151, 156, 158, 164
Club of Rome, 15, 221, 246–7
cockroaches, 298–9
coexistence, 216–19
Cohen, Stanley N., 181
communication: linguistic, and information, 260–5; *see also* language
competition, 216, 217, 220–7
computer feedback systems, 29
conforming strategy (S_+), 26–7, 35, 42–3, 44–6, 55–60, 218, 224
conservation, *see* invariance; parity Conservation
conservative structures, 71–7; and morphogenesis, 72–4, 78, 83, 96–102
contrary strategy (S_-), 26–7, 35, 42–5, 60
Conway, John Horton, 190–6, 217, 297 *n.*
cooperation: in Ehrenfest model, 71–2; in "Equilibrium," 35, 37; games based on, 17, 61; in "Once and for All," 46; in "Random Walk," 32
corpus callosum, 264
Coulomb, Charles Augustin de, 20
Cowan, Jack, 100
Crick, Francis, 164, 303
crystals: of protein molecules and viruses, 74; symmetry and, 103–7 *passim*, 170
cyclical hierarchies, 136–7

Dalton, John, 142
Darwin, Charles, 49, 53–4, 60–1, 164
Darwinian principle of selection, 49–61, 164–5, 256
Davis, Morton, 12, 15
death rate, *see* population size; strategy "decision trees," 7–10, 268–70
decomposition, rate of, 202–4, 208

deduction vs. induction, 299–302
deep structures, 269, 314
Delbruck, Max, 88, 249
Demeny, Paul, 213
Democritus, 112
destabilization, 51, 65
deviation from equal distribution, 38–42
dice, 5, 22–3; musical (Mozart's),
   325–7; in statistical bead games, 30;
   see also bead games
Dirac, Paul, 114, 117
dissipative patterns, 69, 83, 96–102
distribution, probability and (in bead
   games), 33–4, 38–42, 45
DNA (desoxyribonucleic acid), 273–4
dodecahedron, 106–7, 110
"dog flea model," *see* Ehrenfest urn
   model
Doty, Paul, 273
drift, 56, 65, 162
Dürrenmatt, Friedrich, xiv, 176–7, 184,
   278–80

Eccles, John C., 251–5, 264–6, 307
ecological reactions, 91–5
economics, thermodynamics and, 237–9
Edelman, Gerald, 293
Ehrenfest, Paul and Tatyana, 35
Ehrenfest urn model, 34–42, 47–8, 50,
   71–2, 80, 152, 154–5, 163; inverse
   variation of, 44–8
Eigen, Manfred, 179, 309 *n.*
Einstein, Albert, 22, 84, 114, 166
"Elective Affinity" (bead game), 85–9
El-tab el-siga (game), 85
energy, limits on, 244–6
entropy, 101, 141–52, 155, 158–60, 261;
   bead game simulation of, 147–8
enzymes, 60, 90, 98, 159, 160; in
   evolution experiments, 55, 240–2; and
   genetic information, 273–6;

restriction, 181–5; symmetry and,
   123, 128–30; as "teleonomic
   structures," 125
equilibrium: chemical, 81; definition of,
   43, 152–3; statistical, 24, 25, 170, *see
   also* Ehrenfest urn model;
   thermodynamic, 152–60
"Equilibrium" (bead game), 35–43
ergodicity, 170
Eschenmoser, Albert, 303
Escher, Maurits Cornelius, 74, 76,
   105–7, 127
Euler, Leonhard, 138
evolution, 29, 162–4, 243–4, 257;
   chance and, 162–3, 168–9; games of,
   283–9; laws of growth and, 228–35;
   "learning process" of, 21; *see also*
   selection
evolution reactor, 240–2
existentialism, 166, 167
experimentation, science and, 112,
   302–5
exponential growth, 200, 202–9,
   212–14, 216, 218, 220–8, 234

feedback, 29, 90–1
Fermat, Pierre de, 138
Feynman, Richard, 20–1
Fisher, Ronald Aylmer, 54, 263
fission, nuclear, 45–7
"fittest," definition of, 54–7, 59,
   300
fluctuation rectifier, 159–60, 162
fluxes and forces, 240–4
form, concept of, 69–71
formation, rate of, 202–4, 208
Forrester, Jay W., 246–7
Frisch, Max, 258
Frisch, Otto, 46
Fuchs, Walter R., 10
Fucks, Wilhelm, 268, 310–14

fusion, genetic, 183–4
Fux, Johann Joseph, 317, 322, 326

Galileo Galilei, xii, 302–3
game theory, 10–17; and economics, 239
Gardner, Martin, 193
Gauss, Carl Friedrich, 40, 113–14, 138
Gaussian distribution, 40
GEASCOP program, 323
Gell-Mann, Murray, 267
Geminiani, Francesco, 328
generative grammar, 268–70, 280, 282
genetic drift, 56
genetic language, 187–8, 254, 256, 271–7, 280–2; *see also* morphogenesis
genetic manipulation, 180–6
Gentner, Wolfgang, 246
Gerisch, Gunther, 80, 99
Gestalt, concept of, 69–71
Gibbs, Josiah Willard, 164
Gierer, Alfred, 80, 82–3, 100
Gilbert, Walter, 181
global limitations, 221, 224–6, 234
Go (bead game), 16, 64, 205–7, 236
Go-bang (bead game), 205–7
God, existence of, 196
Goethe, Johann Wolfgang von, 74–7, 101, 173, 329
Graun, Karl Heinrich, 314–15
Grimaldi, Francesco, 303
"Growth" (bead game), 209–12
growth, laws of, 200–35

Habermas, Jürgen, 263–4
Hahn, Otto, 46
Haldane, John B. S., 54
"handedness," *see* chirality
Händel, Georg Friedrich, 314–15
harmony, laws of, 313–18

Hartline, Haldan Keffer, 297
Hartmann, Nicolai, 252
Hassenstein, Bernhard, 17–18
Haydn, Joseph, 325
Hayes, W., 280–1
Hayward, R. W., 118–19
"heads or tails," 32–4
Heisenberg, Werner, 20, 103, 117, 309 *n.*
hemoglobin, 125–8, 130
Hess, Benno, 97, 98, 99
Hesse, Hermann, 4–5, 324–5
Hilbert, David, 113
Hiller, Lejaren A., 312, 322, 323
Hilschmann, Norbert, 293
Hindemith, Paul, 317
Hintergarten Study Group, 309 *n.*
Hirsch, Eike Christian, 196
Hoffmann, Geoffrey, 294
Hoppes, D. D., 118–19
Hudson, R. P., 118–19
Huizinga, Johan, xii, 18, 141, 329–30
Hummel, J. J., 325
Humphreys, W. J., 104
Hund, Friedrich, 155
hydrozoans, 80
hyperbolic growth, 201–3, 208, 209, 212, 216, 220, 223
"Hypercycle" (bead game), 229–35

immune systems, 79, 292–7
indifferent population behavior, 27–9
indifferent strategy (S_0), 26–7, 32, 33, 35, 42, 45, 162, 218–19
induction, deduction vs., 299–302
information: aesthetic, theory of, 306–21; creation of, 277–80; entropy as measure of, 144; genetic, 187–8, 254, 256, 271–7, 280–2; linguistic communication and, 260–6; perfect or complete, 11, 12, 14

"Information" (bead game), 290–2
information theory, 148–50, 261, 263
initial conditions, 22–3, 151
instability vs. stability, 161
Institute for Advanced Study,
 Princeton, 12
insurance, probability and, 30–2, 143
internal-combustion engine, 125, 126,
 129
interpretation, value of, 21
interval frequency, 309–13
invariance, 114–17, 151; *see also*
 symmetry
Isaacson, Leonard M., 322

Jacob, François, 164
Jacobson, Roman, 268
Jerne, Niels, 294
Joyce, James, 267

Kant, Immanuel, 196, 263–4, 268
"kemari" (game), 17
Kendren, John, 127, 303–4
Khorana, Ghobind, 303
Kirnberger, Johann Philipp, 325, 328
Klug, Aaron, 75
Köhler, Wolfgang, 102
Kolmogorov, Andrei Nikolaievich, 263
Kornberg, Arthur, 273, 276
Koshland, Daniel, 129
Kupper, Hubert, 322
Kuppers, Bernd, 274–5

Landau, Lev, 72
Landsteiner, Karl, 79
language, 256–7, 259–82; genetic,
 187–8, 254, 256, 271–7, 280–2; of
 immune memory, 292–7; *see also*
 information

large numbers, law of, 19, 23–4, 41–2
Laue, Max von, 127
"learning" networks, 292–7
Le Châtelier's principle, 237–8, 240
Lederberg, Joshua, 183
Lee, Tsung Dao, 115–16, 302
Legendre, Adrien Marie, 138
Leibniz, Gottfried Wilhelm von, 165
Lenin, V. I., 135
Lenneberg, Eric, 268
Leonardo da Vinci, 324
Lietzmann, Walter, 301
"Life" (bead game), 190–6, 217, 236
life, "artificial," 179–80
"life" and "death," game of, 24–9
Ligeti, György, 319, 320
"Limitation of Growth" (bead game),
 224–8
linear growth, 200–5, 209, 216–19, 234
Lipmann, Fritz, 274
lithium, 304–5
Lorentz, Hendrik Antoon, 114
Lorenz, Konrad, 136, 137, 252–4
Loschmidt, Joseph, 152, 155, 159, 160
Lotka, Alfred J., 91, 93–6
Lüders, Gerhart, 116
Lynen, Feodor, 274

macrocosm, definition of, 19
macromolecules, 59; chirality of, 121–4;
 phenotypical variety among, 169; *see
 also* molecules
Malsburg, Christoph von der, 100, 297
 n.
Marin, Thomas, 105, 130, 329
Marxists, 135, 171
mass action, law of, 73, 81
Max Planck Institute (Tübingen), 80,
 99
Max Planck Institute for Behavioral
 Physiology (Seewiesen), 90, 136

Max Planck Institute for Nutritional
Physiology (Dortmund), 97
Maxwell, James Clark, 21, 143, 158–60,
164
McHose, Allen Irvine, 314–16
Meadows, Dennis and Donella, 221,
246–7
Meinhardt, Hans, 80, 82–3, 100
Meitner, Lise, 46
Meixner, Josef, 154
Mellart, James, 106
memory, 164, 254–56, 262; of immune
system, 292–7
Mendel, Gregor Johann, 256
Menninger, Karl, 132, 150 *n.*
Menuhin, Yehudi, 309 *n.*
Mesarović, Mihailo, 244, 246–7
Meselson, Matthew, 181
Michelson-Morley experiment, 302
microcosm, definition of, 19
minimax theorem, 14
Minkowski, Hermann, 114
modulations, 317
molecules: in chemical reactions, 84;
chirality of, 121–4; "intelligent," 27;
reproduction at level of, 57–8;
unpredictability at level of, 19–21; *see
also* genetic language; protein
molecule
Moles, Abraham A., 149–50, 307, 308
Monod, Jacques, xii, 125, 127, 159,
160, 162, 163, 166–9, 277, 289
Morgenstern, Oskar, 12, 15–16
morphogenesis, 77–101, 164
Moulton, William G., 267–8
Mozart, Wolfgang Amadeus, 325, 326,
328
"Mozart's Musical Game," 325–7
music, 309–28
mutation, 78, 163, 168, 256, 300; in
"Selection," 52, 53, 59
Muxfeldt, Hans, 303

natural selection, *see* selection
Neo-Darwinists, 54, 56
Neoplatonists, 70
nerve cells, 29, 100, 254
neutrons, splitting of, 45–6
Newton, Isaac, 142–3
"Nim" (bead game), 10–12
nuclear energy, 245–6
nuclear fission, 45–7
nucleic acids, 59, 60, 221, 229; in
evolution experiments, 55, 169,
240–2; in gene transplants, 181–3;
language of, 272–7
numbers, order of, 136–40

"Once and for All" (bead game), 44–6,
48, 51, 124, 220, 248; *see also*
hyperbolic growth
Onsager, Lars, 154
optimal strategy, 10–15
order, 131–72; of life, 157–72; of
matter, 141–57; of numbers, 136–40;
social, justice and, 131–6, 139, 172
oscillation, 95–7, 153–4

pain centers, 165, 197–8
Parcheesi, 7, 84, 89
parity conservation, 115–19, 156
Parmenides, 120
partial orders, 137–40
partition diagrams, 138–40, 149
Pasteur, Louis, 120–1
Pauli, Wolfgang, 116, 117
Pauling, Linus, 121, 292
pay-off matrix, 12–14; in game of "life"
and "death," 27–9
pecking orders, 136–7
periodic reactions, 95
Perutz, Max, 73, 127, 303–4
Pestel, Eduard, 244, 246–7

pheromones, 42, 90
Philoponos, J., 70
Picht, Georg, 196, 309 *n*.
Planck, Max, 22, 237
plasmids, 183–5
Plato, 107–12
pleasure centers, 165, 197–8
Plutarch, 301
Pohl, Robert Wichard, 165
Poincaré recurrence times, 170
polyp hydra, freshwater, 80, 82–3
Popper, Karl, xiii–xiv, 21–3, 251–4,
  278, 298, 299, 302, 304, 307
population size: biological regulation of,
  43–4; game of "life" and "death"
  and, 24–9; growth in, 199–202, 204,
  208, 213–15, 247–8
Porter, Rodney, 293
Prelog, Vladimir, 303
Prigogine, Ilya, 96, 155–6, 159–60, 240
probability: Darwin's principle and, 56;
  and distribution, in statistical bead
  games, 38–42; in "heads or tails"
  situation, 31–4; *see also* uncertainty
  principle
protein molecule, 24–5, 51, 59, 72–4,
  166; symmetry and, 124–30
proteins, 169, 229; alphabet of, 271–2,
  274; asymmetrical carbon atoms in,
  121–4; and genetic language, 271–6

quantum mechanics, 20, 22, 152, 171–2

Rameau, Jean Philippe, 310, 313–18,
  326, 328
Ramón y Cajal, 295
"Random Walk" (bead game) 32–5,
  39–42, 46, 48, 51, 58, 65, 84, 87, 89
Raphael, Max, 171
rate functions, 200–3

Ratliff, Floyd, 297
reaction games, 81–90
receptors, cell, 79, 82–3, 90
Rechenberg, Ingo, 196
regular solids, 107–11
regulating mechanisms, 29
Reichardt, Werner, 297
relative proximity, 38
relativity theory, 114, 152, 156
relaxation processes, 154
Renyi, Alfréd, 277
reproduction: autocatalytic, 45, 57, 64;
  at molecular level, 57–8; in
  "Selection," 52; *see also* population
  size
reproduction games, 108–9
restriction enzymes, 181–5
reverse transcriptase, 273
Riccioli, Giovanni Battista, 303
Richter, Peter, 294
Riemann, G.F. Bernhard, 114
Rilke, Rainer Maria, 306
RNA (ribonucleic acid), 273–6
RNA game, 284–9
Ruch, Ernst, 139–40
Russell, Bertrand, 269
Rutherford, Ernest, 304

Samuelson, Paul A., 237
Sanger, Frederic, 303–4
Sartre, Jean-Paul, 166, 171
Schiller, Friedrich von, 67
Schillinger, Joseph, 321
Schlogl, Reinhard, 240
Schneider, Dietrich, 90
Schönberg, Arnold, 309, 312, 313, 317,
  318–21, 323, 328
Schopenhauer, Arthur, 153
Schou, Mogens, 304–5
Schrödinger, Erwin, 1, 22, 257
Schubert, Franz Peter, 323

Schuster, Peter, 233
Schwinger, Julian, 116
Scrabble, 289
selection, 49–61, 78, 162–5, 220, 256;
    population growth and, 221–4;
    symmetry and, 120–30; *see also*
    evolution
"Selection" (bead game), 52–3, 55, 56,
    59, 61, 124, 162, 163, 208, 231, 243
self-organization, 29, 91, 101, 164, 169,
    171, 254; of molecules, 60; of
    networks, 297
self-regulating mechanisms, 204
self-reproducing automaton, 189, 192,
    197
self-reproduction, 29, 200, 220–1
sets, 136–8
Shakespeare, William, 267
Shannon, Claude, 148, 149, 261, 263,
    277, 292, 308
Shaw, George Bernard, 267
silkworm, 90
skat, 16–17
slime moulds, 80, 99
snow crystals, 104, 105, 170
socialism, 131–6
solids, Platonic, 107–11
solitaire, 17
Spender, Stephen, 133
Sperry, Roger W., 264
Spiegelman, Sol, 273
Spinoza, Baruch, 166
stability vs. instability, 161
stabilization, 51, 65
stable population behavior, 27–9
Stalin, Joseph, 135
stationary states, 43, 161, 202, 217, 240
statistical physics, 61, 105 *n.*
"Stay Out of 2D" (bead game), 88–90
Stent, Gunter, 180
Stephen, Rudolf, 313, 318
Stockhausen, Karlheinz, 319

stone-paper-scissors (game), 12, 13
Strassman, Fritz, 46
strategy: conforming (S_+), 26–7, 35,
    42–6, 55–60, 218, 224; contrary (S_-),
    26–7, 35, 42–5, 60; indifferent (S_0),
    26–7, 32, 33, 35, 42, 45, 162, 218–19;
    optimal, 10–15
"Struggle" (bead game), 91–5
Sumper, Manfred, 274–5
superexponential growth, 212–14, 234;
    see also hyperbolic growth
"Survival" (bead game), 61–5
symmetry, 103–30; basic types of,
    105–6; broken, 113–24; of
    equilibrium, 153–4, 157; functionality
    and, 120, 125–30; Platonic school
    and, 107–12; selection and, 120–30;
    in Ulam's reproduction game, 108–9
Szilard, Leo, 159

Tarski, Alfred, 271
Teilhard de Chardin, Pierre, 258, 284
Telemann, Georg Philipp, 314–15
teleonomic structures, 125
Temin, Howard, 273
temperature, entropy and, 143–4,
    146–7
Theaetetus, 107
thermodynamics, 142, 145, 152–7,
    237–8; economics and, 237–9; first
    law of, 151; second law of, 151–2,
    155–9, 167–8
Thom, René, 23, 96
tic-tac-toe, 7
time reversal, 151–2, 155–7, 159
topology, differential, 23
transformation games, 91
transplantation, gene, 180–6
triangles, Plato's, 110–12
Turing, Alan M., 96, 188–9, 194
Turing machine, 188–9, 192, 197–8

Tutunkhamen, 84–5
"Twenty Questions" (game), 50

Ulam, Stanislav, 108–9, 189
uncertainty principle, 20–3, 25, 143
unstable population behavior, 27–9
uranium, 45–7; U-238, 47

van't Hoff, Jacobus Hendricus, 121
variable population behavior, 27–9
virus particles, 72–5, 78
Volterra, Vito, 91, 93–6
von Neumann, John, 12, 14, 57, 178, 189–92, 197

Wagner, Carl, 237
Watson, James, 303
Waugh, John S., 156
wave theory, 20, 22
Webern, Anton von, 310–12, 318
Weizsäcker, Carl Friedrich von, 157, 271, 309 *n.*
Weyl, Hermann, 103, 105, 107

Wheatstone, C., 325
Wheeler, John Archibald, 151
Wiener, Norbert, 263
Wigner, Eugene, 141, 163
Wilkins, Maurice, 303
Wilson, Hugh R., 100, 297
Winkler, Ruthild, 179
Wittgenstein, Ludwig, 22, 112, 269
Wolff, Karl-Dietrich, 133
Woodward, Robert, 303
wordplay, 267
World Health Organization, 214
Wright, Sewell, 54
Wu, Chien Shuing, 116, 118–19, 302
Wyman, Jeffries, 127

Xenakis, Yannis, 320

Yamofsky, Charles, 303–4
Yang, Chen Ning, 115–16, 302
Young, Alfred, 138, 140

zero-sum games, 10–15

The Princeton Science Library

Edwin Abbott Abbott	**Flatland: A Romance in Many Dimensions** With a new introduction by Thomas Banchoff
Friedrich G. Barth	**Insects and Flowers: The Biology of a Partnership** Updated by the author
Marston Bates	**The Nature of Natural History** With a new introduction by Henry Horn
John Bonner	**The Evolution of Culture in Animals**
A. J. Cain	**Animal Species and Their Evolution** With a new afterword by the author
Paul Colinvaux	**Why Big Fierce Animals Are Rare**
Peter J. Collings	**Liquid Crystals: Nature's Delicate Phase of Matter**
Pierre Duhem	**The Aim and Structure of Physical Theory** With a new introduction by Jules Vuillemin
Manfred Eigen & Ruthild Winkler	**Laws of the Game: How the Principles of Nature Govern Chance**
Albert Einstein	**The Meaning of Relativity** Fifth edition
Niles Eldredge	**Time Frames: The Evolution of Punctuated Equilibria**
Richard P. Feynman	**QED: The Strange Theory of Light**
J. E. Gordon	**The New Science of Strong Materials, or Why You Don't Fall through the Floor**
Richard L. Gregory	**Eye and Brain: The Psychology of Seeing** Revised, with a new introduction by the author
J.B.S. Haldane	**The Causes of Evolution** With a new preface and afterword by Egbert G. Leigh
Werner Heisenberg	**Encounters with Einstein, and Other Essays on People, Places, and Particles**
François Jacob	**The Logic of Life: A History of Heredity**
Rudolf Kippenhahn	**100 Billion Suns: The Birth, Life, and Death of the Stars** With a new afterword by the author
Hans Lauwerier	**Fractals: Endlessly Repeated Geometrical Figures**
John Napier	**Hands** Revised by Russell H. Tuttle

J. Robert Oppenheimer	**Atom and Void: Essays on Science and Community** With a preface by Freeman J. Dyson
John Polkinghorne	**The Quantum World**
G. Polya	**How to Solve It: A New Aspect of Mathematical Method**
Hazel Rossotti	**Colour, or Why the World Isn't Grey**
David Ruelle	**Chance and Chaos**
Henry Stommel	**A View of the Sea: A Discussion between a Chief Engineer and an Oceanographer about the Machinery of the Ocean Circulation**
Hermann Weyl	**Symmetry**